T0345158

INFORMATION SECURITY GOVERNANCE

INFORMATION SECURITY GOVERNANCE

A Practical Development and Implementation Approach

KRAG BROTBY

A JOHN WILEY & SONS, INC., PUBLICATION

Published by John Wiley & Sons, Inc., Hoboken, New Jersey.
Published simultaneously in Canada.

Library of Congress Cataloging-in-Publication Data:

Brotby, W. Krag.
Information security governance : a practical development and implementation approach / Krag Brotby.
p. cm. — (Wiley series in systems engineering and management)
Includes bibliographical references and index.
ISBN 978-0-470-13118-3 (cloth)
1. Data protection. 2. Computer security—Management. 3. Information technology—Security measures. I. Title.
HF5548.37.B76 2009
658.4'78—dc22 2009007434

Printed in the United States of America.

10 9 8 7 6 5 4 3 2 1

Contents

Acknowledgments

A debt of gratitude is acknowledged to my wife, Melody, who graciously accepted many late night hours in assembling this work. Also acknowledged are those who have supported this effort by giving their time to advise, review, and comment on this exposition, including ISACA associates of notable competence Bruce Wilkins, Gary Barnes, and Ron Hale; other professionals including Charles Neal, formerly of the FBI, and Adam Hunt, currently with Inland Revenue in New Zealand. Thanks are also due to John Sherwood and David Lynas from the United Kingdom for their assistance and support with the SABSA architectural material in this work. And finally, appreciation is due to my unusually helpful and cooperative publisher and staff in bringing this hopefully illuminating work to light.

K. B.

Introduction

For most organizations, reliance on information and the systems that process, transport, and store it, has become absolute. In many organizations, information *is* the business. Actionable information is the basis of knowledge and as Peter Drucker stated over a decade ago, "Knowledge is fast becoming the sole factor of productivity, sidelining both capital and labor."*

This notion is buttressed by recent studies showing that over 90% of organizations that lose their information assets do not survive. Research also shows that currently, information assets and other intangibles comprise more than 80% of the value of the typical organization.

Yet, even as this realization has belatedly started to reach executive management and the boardroom in recent years, organizations are plagued by evermore spectacular security failures and losses continue to mount. This is despite a dramatic a rise in overall spending on a variety of security- or assurance-related functions and national governments imposing a host of increasingly restrictive regulations.

This host of new security-related regulations has in turn led to a proliferation of the number and types of "assurance" functions. Until recently, for example, "privacy" officers were unheard of, as were "compliance" officers. Now, they and others, such as the Chief Information Security Officer, are commonplace. It should be noted that all assurance functions are an aspect of what is arbitrarily labeled "security" and, indeed, what is called "security" is invariably an assurance function. In turn, both are elements of risk management.

Not only has the diversity of "assurance" functions increased, the requirements for these activities in many of an organization's other operations are now the norm. Examples include the HIPAA "privacy assurance" functions generally handled by Human Resources, or SOX disclosure compliance as a purview of Finance.

For many larger organizations, a list of assurance-related functions might include:

- Risk management
- BCP/DR
- Project office
- Legal

*Drucker, Peter; *Management Challenges for the 21st Century, Harpers Business,* 1993.

- Compliance
- CIO
- CISO
- IT security
- CSO
- CTO
- Insurance
- Training/awareness
- Quality control/assurance
- Audit
- HR
- Privacy

Combined, these assurance functions constitute a considerable percentage of an organizations' operating budget. Yet, ironically, this increase in assurance functions has in many organizations led to a *decrease* in "safety" or security. This is a consequence of increasingly fragmenting assurance functions into numerous vertical "stovepipes" only coincidentally related to each other and to the organization's primary business objectives. This, despite the fact that all of these activities serve fundamentally only one common purpose: *the preservation of the organization and its ability to continue to operate and generate revenue.*

To compound the problem, these functions invariably have different reporting structures, often exist in relative isolation, speak different languages, and more often than not operate at cross purposes. Typically, they have evolved over a period of time, usually in response to either a crisis du jour or to mounting external regulatory pressures. Their evolution has often involved arbitrary factors unrelated to improving security functionality, efficiency, or effectiveness.

As these specialized assurance functions have developed, national or global associations have formed to promote the specialty. One outcome of this "specialty"-centric perspective has been to widen the divide between elements of what should arguably be a continuous "assurance" process, seamlessly dovetailed and aligned with the business.

So what is the way forward? It has become increasingly clear that the solution lies in elevating the governance of the typical myriad assurance functions to the highest levels of the organization. Then, as with other critical, expensive organizational activities, an assurance governance framework must be developed that will integrate these functions under a common strategy tightly aligned with and supporting business objectives.

Alternatively, for most organizations, failure to implement effective information security governance will result in the continued chaotic, increasingly expensive, and marginally effective firefighting mode of operation typical of most security departments today. Tactical point solutions will continue to be deployed, and effective administration of security and integration of assurance functions will have no impetus and remain merely a concept in the typically fragmented multitude of "assurance-" and security-related stovepipes. Allocation of security resources is likely to remain haphazard and unrelated to risks and impacts as well as to cost-effectiveness. Breaches and losses will continue to grow and regulatory compliance will be more costly to address. It is clear that senior management will increasingly be seen as responsible and legally liable for failing the requirements of due care and diligence. Customers will demand greater care and, failing to get it, will vote with their feet, and the correlation between security, customer satisfaction, and business success will become increasingly obvious and reflected in share value.

Against this backdrop, this book provides a practical basis and the tools for developing a business case for information security (or assurance) governance, developing and implementing a strategy to increasingly integrate assurance functions over time, improving security, lowering costs, reducing losses, and helping to ensure the preservation of the organization and its ability to operate.

Chapters 1 through 6 provide the background, rationale, and basis for developing governance. Chapters 7 through 14 provide the tools and an approach to developing a governance implementation strategy.

Developing a strategy for governance implementation will, at a high level, consist of the following steps:

1. Define and enumerate the desired outcomes for the information security program
2. Determine the objectives necessary to achieve those outcomes
3. Describe the attributes and characteristics of the desired state of security
4. Describe the attributes and characteristics of the current state of security
5. Perform a comprehensive gap analysis of the requirements to move from the current state to the desired state of security
6. Determine available resources and constraints
7. Develop a strategy and roadmap to address the gaps, using available resources within existing constraints
8. Develop control objectives and controls in support of strategy
9. Create metrics and monitoring processes to:
   - Measure progress and guide implementation
   - Provide management and operational information for decision support

Chapter 1

Governance Overview—How Do We Do It? What Do We Get Out of It?

1.1 WHAT IS IT?

Governance is simply the act of governing. The *Oxford English Dictionary* defines it as "The act or manner of governing, of exercising control or authority over the actions of subjects; a system of regulations."

The relevance of governance to security is not altogether obvious and most managers are still in the dark about the subject. Information security is often seen as fundamentally a technical exercise, purely the purview of information technology (IT). In these cases, the information security manager generally reports directly or indirectly to the CIO but in some cases may report to the CFO or, unfortunately, even to Operations.

In recent years, there has also been an increase in the number of senior risk managers, or CROs, and, in some cases, Information Security reports through that office. Although these organizational structures often work reasonably well in practice, provided the purview of security is primarily technical and the manager is educated in the subject and has considerable influence, in many cases they do not work well and, in any event, these reporting arrangements are fundamentally and structurally deficient. This contention is often subject to considerable controversy even among security professionals. However, analysis of the wide range of activities that must be managed for security to be effective and study of the best security management shows that it requires the scope and authority equivalent to that of any other senior manager. To be effective, security and other assurance activities are regulatory functions and cannot report to the regulated without creating an untenable structural conflict of interest. Maintaining a distinction between regulatory and operational functions is critical, as each has a very different focus and responsibility. The former is related to safety and the latter to performance, and it is not unusual for tension to exist between them.

Part of the reason that the requirement for separation of security from operational activities is not evident is that the definitions and objectives of security generally lack clarity. Asking the typical security manager what the meaning of security is will elicit the shop-worn response of "ensuring the confidentiality, integrity, and availability of information assets." Pointing out that that is what it is supposed to do, that is its mission, and not what it is, generally elicits a blank stare. Probing further into the objectives of security will usually result in the same answer.

The lack of clarity about what security should specifically provide, how much of it is enough, and knowing when that has been achieved poses a problem and contributes to the confusion over the appropriate organizational structure for security. Lacking clear objectives, a definition of success, and metrics about when it has been achieved begs the question, *What does a security manager actually do*? How is the manager to know when he or she is managing appropriately? What is his or her performance based on? How does anyone know?

In other words, as in any other business endeavor, we manage for defined objectives, for outcomes. Objectives define intent and direction. Performance is based on achieving the objectives. Metrics determine whether or not objectives are being achieved.

1.2 BACK TO BASICS

If there is a lack of clarity looking ahead, reverting to basics may help shed light on the subject. Security fundamentally means safety, or the absence of danger. So in fact, IT or information security is an assurance function, that is, it provides a level of assurance of the safety of IT or information. Of course, it must be recognized that the safety of an organization's information assets typically goes a considerable distance beyond the purview of IT.

IT is by definition technology centric. IT security is by definition the security related to the technology. From a business or management perspective, or, indeed, from a high-level architectural viewpoint, IT is simply a set of mechanisms to process, transport, and store data. Whether this is done by automated machinery or by human processes is not relevant to the value or usefulness of the resultant activities. It should be obvious, therefore, that IT security cannot address the broader issue of information "safety."

Information security (IS) goes further in that it is information centric and is concerned with the "payload," not the method by which it is handled. Studies have clearly shown that the risks of compromise are often greater from the theft of paper than from IT systems being hacked. The loss of sensitive and protected information is five times greater from the theft or loss of laptops and backup tapes than it is from being hacked. These are issues typically outside the scope of IT security. The fact that the information on these purloined laptops or tapes is infrequently encrypted is not a technology problem either; it is a governance and, therefore, a management problem.

To address the issues of "safety," the scope of information security governance must be considerably broader than either IT security or IS. It must endeavor to initiate a process to integrate the host of functions that in the typical organization are related to the "safety" of the organization. A number of these were mentioned in the Introduction, including:

- Risk management
- BCP/DR
- Project office
- Legal
- Compliance
- CIO
- CISO
- IT security
- CSO
- CTO
- CRO
- Insurance
- Training/awareness
- Quality control/assurance
- Audit

To this list we can add privacy and, perhaps more importantly, facilities. Why facilities? Consider the risks to information "safety" that can occur as a function of how the facility operates: the physical security issues, access controls, fire protection, earthquake safety, air-conditioning, power, telephone, and so on. Yet, risk assessments in most organizations frequently do not consider these elements.

The advantage of using the term "organizational safety" and considering the elements required to "preserve" the organization is that the task of security management becomes clearer. It also becomes obvious that many of the other "assurance" functions that deal with aspects of "safety" must be somehow integrated into the governance framework. It also becomes clear that most attempts to determine risk are woefully inadequate in that they fail to consider the broad array of threats and vulnerabilities that lie beyond IT and, indeed, beyond IS as well.

1.3 ORIGINS OF GOVERNANCE

It may be helpful to consider how the whole issue of governance arose to begin with to understand its relevance to information security. The first instance of the appearance of corporate governance seems to be due to economist and Noble laureate Milton Friedman, who contended that "Corporate Governance is to conduct the business in accordance with owner or shareholders' desires, while conforming to the basic rules of the society embodied in law and ethical custom." This definition was based on his views and the economic concept of market value maximization that underpins shareholder capitalism.

The basis for modern corporate governance is probably a result of the Watergate scandal in the United States during the 1970s, which involved then President Nixon ordering a burglary of the opposition party's headquarters. The ensuing investigations by U.S. regulatory and legislative bodies highlighted organizational control failures that allowed major corporations to make illegal political contributions and to bribe government officials. This led to passage of the U.S. Foreign and Corrupt Practices Act of 1977 that contained specific provisions regarding the establishment, maintenance, and review of systems of internal control

In 1979, the U.S. Securities and Exchange Commission proposed mandatory reporting on internal financial controls. Then, in 1985, after the savings and loan collapse in the United States as a result of aggressive lending, corruption, and poor bookkeeping, among other things, the Treadway Commission was formed to identify main causes of misrepresentation in financial reports and make recommendations. The 1987 Treadway Report highlighted the need for proper control environments, independent audit committees, and objective internal audit functions. It suggested that companies report on the effectiveness of internal controls and that sponsoring organizations develop an integrated set of internal control criteria.

This was followed by the Committee of Sponsoring Organizations (COSO), which was formed and developed the 1992 report stipulating a control framework that was endorsed and refined in four subsequent U.K. reports: Cadbury, Rutteman, Hampel, and Turnbull.

Scandals and corporate collapses in the United Kingdom in the late 1980s and early 1990s led the government to recognize that existing legislation and self-regulation were not working. Companies such as Polly Peck, British & Commonwealth, BCCI, and Robert Maxwell's Mirror Group News International in United Kingdom were some of the high-profile victims of the irrational exuberance of the 1980s and were determined to be primarily a result of poor business practices.

In 1991, the Cadbury Committee drafted a code of practices defining and applying internal controls to limit exposure to financial loss.

Subsequent to the most spectacular failures in recent times of Enron, Worldcom, and numerous other companies in the United States, the draconian Sarbanes–Oxley Act of 2002 required financial disclosure, testing of controls and attestation of their effectiveness, board-level financial oversight, and a number of other stringent control requirements.

In January 2005, the Bank of England, the Treasury, and the Financial Services Authority in the United Kingdom published a joint paper on supervisory convergence addressing many of the same issues as Sarbanes–Oxley.

Currently, the global revolution in high-profile governance regulation has resulted in the following, among others:

Financial Services Authority (U.K.)
Combined Code–Turnbull, Smith, Higgs (U.K.)
Sarbanes–Oxley (U.S.)
OECD Principles of Corporate Governance 1999 (G7)
Russian "Code of Corporate Governance" 2002
World Bank Governance Code of Best Practices (global)
BASEL II Accords (global financial organizations)
HIPPA (medical, U.S.)
Corporations Act 2001 (Australia)

1.4 GOVERNANCE DEFINITION

The Information Security Audit and Control Association (ISACA), a global organization originally formed in the late 1960s as an association of IT auditors and now comprising over 70,000 security professionals states that governance is:

> The set of responsibilities and practices exercised by the board and executive management with the goal of providing *strategic direction,* ensuring that *objectives are achieved,* ascertaining that *risks are managed* appropriately, and verifying that the enterprise's *resources are used responsibly.*

The Organization for Economic Cooperation and Development (OECD) Principles states that governance should include the "*structure* through which the objectives of the enterprise are set, and the *means* of attaining those objectives and *monitoring* performance are determined . . ." [1].

Further reading of this definition finds that it includes:

- Organizational structure
- Strategy (and design)
- Policy and corresponding standards and procedures
- Strategic and operational plans
- Awareness and training
- Risk management
- Controls and countermeasures
- Audits, monitoring, and metrics
- Other assurance activities

1.5 INFORMATION SECURITY GOVERNANCE

Obviously, information security has to address the standard notions of security, which include:

- **Confidentiality**—Information is disclosed only to authorized entities
- **Integrity**—Information has not been subject to unauthorized modification
- **Availability**—Information can be accessed by those that need it when they need it
- **Accountability and Nonrepudiation** are also required for digital commerce.

But to address the broader issue of "safety," the notion of preservation must also be considered:

> It is no longer enough to communicate to the world of stakeholders why we exist and what constitutes success, *we must also communicate how we are going to protect our existence.* [2]

This suggests two specific recommendations for steps to be taken:

1. Develop a strategy for *preservation* alongside a strategy for progress.
2. Create a clearly articulated purpose and preservation statement.

1.6 SIX OUTCOMES OF EFFECTIVE SECURITY GOVERNANCE

Extensive research and analysis by ISACA [3] has determined that effective information security governance should result in six outcomes, including:

1. **Strategic alignment**—aligning security activities with business strategy to support organizational objectives
2. **Risk management**—executing appropriate measures to manage risks and potential impacts to an acceptable level
3. **Business process assurance/convergence**—integrating all relevant assurance processes to maximize the effectiveness and efficiency of security activities
4. **Value delivery**—optimizing investments in support of business objectives
5. **Resource management**—using organizational resources efficiently and effectively
6. **Performance measurement**—monitoring and reporting on security processes to ensure that business objectives are achieved

Defining the specifics of these outcomes for an organization will result in determining governance objectives. A thorough analysis of each of the six will provide a basis for clarifying the requirements and expectations of information security, and, subsequently, the sort of structure and activities needed to achieve those outcomes.

1.7 DEFINING INFORMATION, DATA, AND KNOWLEDGE

Many of the terms used in IS and IT have lost their precise meaning and are often used interchangeably. For the purposes of gaining clarity on the subject of *information* security governance, it is useful to define the terms. The admonition of more than fifty years ago that "A man's judgment cannot be better than the information on which he has based it" [4] is still valid.

Data is the raw material of information. Information, in turn, may be defined as data endowed with relevance and purpose. Knowledge is created from information. Knowledge is, in turn, captured, transported, and stored as organized information. Knowledge is created from actionable information. Peter Drucker recognized the emerging importance of knowledge more than a decade ago, stating, "Knowledge is fast becoming the sole factor of productivity, sidelining both capital and labor" [5].

1.8 VALUE OF INFORMATION

It may be that the very nature of information and the knowledge based on it is so ubiquitous and transparent that we generally fail to recognize its true value and our utter dependence on it. It is, however, abundantly clear that information is the one asset organizations cannot afford to lose; their very existence depends on it.

> Information, the substance of knowledge, is essential to the operation of all organizations and may comprise a significant proportion of the value as well.
>
> Companies may survive the loss of virtually all other assets including people, facilities, and equipment, but few can continue with the loss of their information, i.e., accounting information, operations and process knowledge and information, customer data, etc. [6]

Studies have shown that the information residing in an organization is, in most instances, the single most critical asset. This is demonstrated by an investigation performed by Texas A&M University which showed that [7]:

- 93% of companies that lost their data center for 10 days or more due to a disaster filed for bankruptcy within one year of the disaster.
- 50% of businesses that found themselves without data management for this same time period filed for bankruptcy immediately.

This and other studies of the consequences of loss of information, such as occurred in 1993 to a number of the businesses in the World Trade Center in New York that had all their information stored in the data center in the basement, which was destroyed by a truck bomb, substantiate the dire consequences to organizations that lose use of their information.

Another marker for the value of information beyond survival is a recent study by the Brookings Institution that found that typically, an organization's information and other intangible assets account for more than 80% of its market value [8].

REFERENCES

1. OECD 99, *Principles of Corporate Governance*, 2004.
2. Kiely, L. and T. Benzel, *Systemic Security Management,* Libertas Press, 2006.
3. Information Security And Control Association, *2008 CISM Review Manual.*
4. Chomsky, D., "The Mechanisms of Management Control at the New York Times," *Media, Culture & Society,* Vol. 21, No. 5, 579–599, 1999.
5. Drucker, P., *Management Challenges for the 21st Century,* Butterworth-Heinemann Ltd., Oxford, 1999.
6. Brotby, K., *Information Security Governance: A Guide for Boards of Directors and Senior Management,* IT Governance Institute, 2006.
7. Moskal, E., "Business Continuity Management Post 9/11 Disaster Recovery Methodology," *Disaster Recovery Journal,* Vol. 19, Issue 2, 2006.
8. Osterlund, A., "Decoding Intangibles," *CFO Magazine,* April 2001.

Chapter **2**

Why Governance?

> Information security is not only a technical issue, but also a business and governance challenge that involves risk management, reporting, and accountability. Effective security requires the active engagement of executive management to assess emerging threats and provide strong cyber security leadership. The term penned to describe executive management's engagement is *corporate governance.* Corporate governance consists of the set of policies and internal controls by which organizations, irrespective of size or form, are directed and managed. Information security governance is a subset of organizations' overall governance program. Risk management, reporting, and accountability are central features of these policies and internal controls. [1]

Can a business case be convincingly made to implement information security governance or is it simply another needless layer of complexity designed to boost security department budgets? Although there are relatively few studies, the conclusions provide strong support for the necessity. Combined with the continuing growth of preventable cybercrime, mounting losses, and the all-too-common chaotic, unintegrated state of information security also suggests that there is simply no other rational approach to achieving effective enterprise-wide security given the complexity, breadth, and the sheer number of "moving parts."

One of the more interesting and significant recent studies by the Aberdeen Group found that "Firms operating at best-in-class (security) levels are lowering financial losses to less than one percent of revenue, whereas other organizations are experiencing loss rates that exceed five percent" [2].

To the extent that the research proves accurate, this dramatic finding would appear to make any organization not practicing "best-in-class" security bordering on sheer recklessness and its management utterly failing its responsibilities. For any organization, the results of this study, suggesting that they might lower security-related losses by more than 80%, would seem to make a compelling case for effective security governance to drive "best-in-class" security.

The study involved a number of companies of various sizes, but extrapolating from an organization with $500 million U.S. in revenues, a reduction of losses from

$25 million (5%) to $5 million (1%) annually would fund substantial security efforts and probably leave some money left over.

The question that arises then is what constitutes "best-in-class" security? Some would suggest that it means adherence to so-called "best practices" that are the cornerstone of ITIL. In some cases, however, best practices may be appropriate; in other cases, they may excessive or insufficient. A persuasive argument can be made that "best practices" is merely a substitute for a lack of real knowledge. That is to say, one size will not fit all, and with good planning and effective metrics, adequate and sufficient practices are a far more cost-effective approach. In any event, practices of any sort, whether best or not, must be managed in an integrated manner consistent with supporting business objectives to be of any significant value to an organization.

Although it may not be possible to provide a specific set of precise specifications to define "best in class" or "appropriate" level, there are several internationally recognized and accepted gauges and standards available to assess what that entails. The attributes and characteristics defined in the CobiT version of Capability Maturity Model (CMM) Level 4 paints a clear picture and would fulfill the requirement for most organizations. It states:

4—Managed and Measurable

- The assessment of risk is a standard procedure and exceptions to following the procedure would be noticed by IT management. It is likely that IT risk management is a defined management function with senior level responsibility. Senior management and IT management have determined the levels of risk that the organization will tolerate and have standard measures for risk/return ratios.
- Responsibilities for Information security are clearly assigned, managed and enforced. Information security risk and impact analysis is consistently performed. Security policies and practices are completed with specific security baselines. Security awareness briefings have become mandatory. User identification, authentication and authorization are standardized. Security certification of staff is established. Intrusion testing is a standard and formalized process leading to improvements. Cost/benefit analysis, supporting the implementation of security measures, is increasingly being utilized. Information security processes are coordinated with the overall organization security function. Information security reporting is linked to business objectives.
- Responsibilities and standards for continuous service are enforced. System redundancy practices, including use of high-availability components, are consistently deployed. [3]

Although it is somewhat imprecise and subjective, CMM is an integral part of CobiT [3] and provides a straightforward intuitive approach that most find easy to apply.

A more detailed and specific approach is provided by the ISO/IEC 27002 Code of Practice and the 27001 Standard that specifies comprehensive requirements for governance, implementation, metrics, controls, and compliance.

High-level governance requirements are also set forth comprehensively in FISMA documentation pursuant to the U.S. Federal Information Security Management Act.

Whichever approach is utilized, the objective is to achieve "best-in-class" security through good governance, which, in summary, will ensure:

- Assignment of roles and responsibilities
- Periodic assessments of risks and impact analysis
- Classification and assignment of ownership of information assets
- Adequate, effective, and tested controls
- Integration of security in all organizational processes
- Implementation of processes to monitor security elements
- Effective identity and access management for users and suppliers of information
- Meaningful metrics
- Education of all users, including management and board members, of information security requirements
- Training as needed in the operation of security processes
- Development and testing of plans for continuing the business in case of interruption or disaster

2.1 BENEFITS OF GOOD GOVERNANCE

A number of identifiable benefits will devolve from implementing effective information security governance, depending on the current state of security and particulars of the organization. The following subsections discuss be some of the more direct and obvious benefits but there are likely to be other, less obvious ones. For example, embarking on a program to implement governance as detailed in the following pages is likely to improve the awareness and commitment of management and result in a better "tone at the top." This in turn may initiate a culture more conducive to security.

2.1.1 Aligning Security with Business Objectives

Although it seems an obvious requirement, the majority of organizations globally do not have a program or process to align IT strategy, much less security activities, with the objectives of the business. This was vividly highlighted by the 2006 *Global State of Information Security Governance* study of more than seven thousand organizations by the IT Governance Institute [4]. It revealed that processes to align IT strategy with business strategy had only been implemented by 16% of the respondents. Another 12% indicated that they were in the process of implementing a program to address the issue. The remaining 72% of organizations did not know what guided their IT and security activities (Table 2.1).

Demonstrably aligning security and other "assurance" functions directly and specifically with business strategy and objectives arguably provides a number of

Table 2.1. Implementation of IT strategy by businesses

	Have implemented	Implementing now	Considering implementing	Not considering
IT strategy alignment with business strategy	16%	12%	21%	51%
Resource management	18%	12%	20%	50%
Value delivery	9%	9%	21%	61%
Risk management	9%	9%	16%	66%
Performance of IT	10%	10%	14%	66%
ROI management of IT	7%	8%	13%	72%

benefits. It serves to provide greater support for and cooperation with security efforts by business owners and senior management. This will in turn improve the "tone at the top" and the overall security culture as it counters the perception that "security" is a bottomless cost pit whose main objective is to hinder business activities and complicate life generally.

The ongoing, effective alignment of security with business can only be accomplished by properly implemented information security governance which creates the structure and process to align not only IT but also security with business by defining the objectives, creating the appropriate linkages, and providing an implementation strategy along with suitable metrics to track progress.

Considering that just the limited subset of assurance functions called "security" on average has risen to consume 2% of corporate revenues, it is evident that integration with and support of the business is worthwhile.

2.1.2 Providing the Structure and Framework to Optimize Allocations of Limited Resources

Resources are always limited. A good security governance framework coupled with a well-developed strategy will include processes to identify where the greatest benefit will be derived in terms of supporting business objectives. If, for example, a part of the business strategy is to automate the supply chain using information systems, it is obvious that elements of security such as availability and integrity are critical to a successful implementation. If the strategy calls for developing an online business, confidentiality can be added to the requirements of availability and integrity, and the necessity for security is obvious. Allocating resources for appropriate security to support strategic functions will clearly be of benefit to the business and is a demonstrable example of alignment.

What is perhaps less obvious relates to protecting the organization, that is, making it "safe." Allocation of security resources are virtually *never proportional* to the risks or impacts facing the typical organization. The ISACA Global Security Governance Survey found that only 30% of respondents had processes in place for resource management, only 18% analyzed value delivery, and a mere 15% determined the return on investment of IT.

Absent any processes in place to assess or analyze resource allocation, it is perhaps not surprising that security spending generally appears to be totally unrelated to actual risks, impacts, or maximizing return on investments.

From an in-depth study performed in 2006 by PGP Corporation and Vontu of 31 organizations suffering security breaches [5], it is clear that the allocation of security resources in these, and apparently most, organizations did not correspond to the source of losses. Although the analysis showed the average breach loss was nearly $5 million, the majority did not arise from technical system breaches.

Systems hacked	7%
Malicious insiders or malicious code	9%
Paper records	9%
Compromised, lost, or stolen backups	19%
Third party or outsourcer	21%
Lost or stolen laptops or other devices	35%

It has been argued that IT security was shown to be working by the fact that the smallest percentage of losses resulted from hacking, but it more clearly shows that consistent security baselines do not exist and that resource allocations are not based on risks and impacts

For example, since 35% of the breach losses averaging $4.8 million U.S. were caused by information gained from lost or stolen portable devices, that suggests that an average of $1.7 million could be saved by the simple expedient of encrypting the storage of portable devices. Licensing the appropriate software to accomplish this would typically not amount to more than a few tens of thousands for even a large organization. Clearly the return on investment in this case is staggering.

Security governance addresses the issues of allocation based on what provides the best return on investment, and what the greatest sources of manageable risks are.

2.1.3 Providing Assurance that Critical Decisions are Not Based on Faulty Information

A primary function of information systems is decision support. Guiding the organization and managing operations relies on timely, correct information. Implementing appropriate governance structures provides the processes to provide assurance of information confidentiality, integrity, and availability.

2.1.4 Ensuring Accountability for Safeguarding Critical Assets

In many organizations, roles and responsibilities are not clearly defined. The typical result in the event of security "malfunction" is that it is often unclear who is responsible. This situation is usually compounded by a lack of clear and adequate authority as well. The lack of clarity of responsibility and accountability is often accompanied by a culture of blame, a culture in which mistakes are not used to learn from but as a basis for retribution. From a security or safety standpoint, this is counterproductive at best and downright dangerous at worst.

Governance can counter this situation by making the intent, direction, and expectations of management explicit, clearly defining roles and responsibilities, and creating monitoring processes for assurance of acceptable performance.

2.1.5 Increasing Trust of Customers and Stakeholders

Detailed analysis by PGP Corporation and Vontu of 31 companies that suffered information breaches in 2006 found that losses averaged $4.8 million. And, perhaps more significantly, 19% of customers terminated their relationship with the breached company and another 40% were considering doing so! For most organizations, losing nearly 60% of their customers could be a major problem.

For many organizations, such as financial institutions, stock brokers, and retailers, customer trust, confidence, and satisfaction are the key indicators that correlate with growth of the business. For any business that operates online, there is typically nothing that causes customer dissatisfaction more that compromised accounts or stolen identities, all purviews of security.

2.1.6 Increasing the Company's Worth

Increasingly, institutional and sophisticated investors consider governance in their investment decisions. There are now a number of rating organizations that include various aspects of governance as a key evaluation marker.

McKinsey and Company, in conjunction with Institutional Investors, Inc., published studies in the *McKinsey Quarterly* that concluded that major international investors were willing to pay a premium for shares in a company that is known to be well governed [6]. The premium ranged from 11 to 16 percent in 1996 to 18 to 28 percent in 2000. With the advent of regulations, such as those imposed by Sarbanes–Oxley requiring disclosure of the effectiveness of controls and attestation to the accuracy of financial reporting, these studies suggest obvious implications for adequate and effective security governance.

2.1.7 Reducing Liability for Information Inaccuracy or Lack of Due Care in Protection

There are many chapters yet to be written in jurisprudence regarding information security issues, but some aspects have emerged that must be considered in terms of organizational liability for inaccurate information.

One example is Guess Jeans, sued by the FTC for misrepresenting on their website that their customer information was stored in ureadable encrypted form at all times, which turned out not to be true and resulted in a multimillion dollar fine.

A carefully watched case not yet adjudicated but winding its way through U.S. Federal Court at this writing is CardSystems, a payments processor that was hacked and exposed more than 40 million credit card records. The issues of provable damages, duty of due care, and public policy are being addressed and likely to have significant impact.

ChoicePoint, a supplier of personal information that failed to practice adequate standards of care in 2006 paid a $10 million U.S. fine and another $5 million in stipulated damages.

The most recent case involving TJX and the compromise of more than 46 million credit records has as of this writing cost the company in excess of $250 million U.S., an expensive lesson in bargain basement security.

In the forgoing cases, it is clear that it would have been far less expensive to address the security issues properly.

2.1.8 Increasing Predictability and Reducing Uncertainty of Business Operations

The timeworn statement that business thrives under the rule of law is, of course, timeworn because it is true. A level of predictability is needed to determine pricing, costs, and operations. Risk comes at a cost; the higher the risk, the greater the cost. Whether true or not, the quip, "There's a reason there's no Starbucks in Baghdad" serves to illustrate the point. Business simply does not choose to operate where risks are high and unmanageable.

The implementation of effective governance takes security from a haphazard, usually chaotic operation to a consistent, standardized, integrated, effective business activity.

2.2 A MANAGEMENT PROBLEM

In the past, security was typically the unwanted stepchild of business, relegated to corporate Siberia, and at best an afterthought. As losses mounted and regulators became increasingly insistent on organizations implementing reasonable security measures, grumbling executives grudgingly supported some minimum technical security measures, usually in the form of succumbing to vendors' assertions that a particular box or software would solve the organizations' security issues. Continuing losses and, sometimes, catastrophic experiences has started to convey the realization that security, or safety, will require a concerted effort supported at the highest levels of management. The fact is, no amount of unmanaged, unintegrated technology is likely to make an organization "safe." Security requires the same level of management commitment and resources as any other major organizational activity. Security is a management problem, not a technical problem.

REFERENCES

1. The Corporate Governance Task Force, 2004, http://www.cyberpartnership.org/InfoSec-Gov4_04.pdf.

2. Aberdeen Group, "Best Practices in Security Governance," 2005.
3. Frequently Asked Questions—CobiT, www.isaca.org/Content/Navigation menu/CobiT6/FAQ6.
4. IT Governance Institute, *Global State of Information Security,* 2006.
5. Ponemon Report, "The 2006 Cost of a Data Breach," PGP Corp., Vontu, www.pgp.com/insight/newsroom/press_releases/2006/ponemon.html.
6. McKinsey and Institutional Investors Inc., *McKinsey Quarterly,* "McKinsey/KIOD Survey on Corporate Governance," January 2003, www.mckinsey.com/clientservice/organizationleadership/service/corpgovernance/pdf/cg_survey.pdf.

Chapter 3

Legal and Regulatory Requirements

Reacting to the spectacular failures of security and governance in recent times, governments have enacted a raft of new laws and regulations during the past few years that have had some impact on most organizations and the practices of information security. When these measures proved inadequate to address burgeoning identity theft and fraud, the credit association comprising VISA, Mastercard, Discover, JCB, and American Express instituted PCI (Payment Card Industry) Data Security Standards for all online merchants using credit cards. PCI is of particular interest insofar as it affects any organization accepting credit cards, which is a high percentage globally. It is also of interest that a global standard for security is being imposed by private organizations and is likely to have greater impact than governmental actions.

The other global regulations affecting all international financial organizations is BASEL II, with its requirements for operational risk management and tighter standards for maintaining reserves.

Although these efforts will not address the double-digit growth of phishing and pharming, it is likely to eventually reduce breach losses, which reached around ninety million names in 2005. The subsequent spectacular failure of security at TJX in 2007 resulting in the compromise of over 46 million records by itself and costs to the organization as of this writing exceeding $250 million U.S. is accelerating efforts at enforcement.

Some aspects of security have undoubtedly improved as a result of the more pervasive laws such as Sarbanes–Oxley (SOX), HIPAA, Gramm–Leach–Bliley, BASEL II, EU Privacy Directive, privacy breach notice laws started with California's SB1386, and a number of others. Yet, surprisingly, even with some of the draconian penalties for wanton violators of SOX, recent surveys indicate less than half of affected U.S. organizations are reasonably in compliance or in the process and the rest are not planning to comply. Although this is primarily a function of the current lacklustre compliance enforcement, that may change with the political winds and increased focus on compliance. In the meantime, there are major efforts by corporations to convince the U.S. Congress to reduce the requirements of these regulations, which they consider onerous and excessive.

While they may be some shuffling of requirements, the trend toward greater and more restrictive regulation will continue. IT systems have become so pervasive and critical that nations cannot afford for them to fail significantly. Research shows that about 80% of national critical infrastructure is in private hands and universally subject to disruption and security-related failures.

Despite the avalanche of regulatory requirements, compliance remains surprisingly low at this writing. A 2006 study by Price Waterhouse Coopers (PWC) [1] in 50 countries with over 7700 respondents found the states of compliance with current standards and regulations as a percentage of organizations admitting that they need to be in compliance with specific laws but are not, as listed in Table 3.1.

There are undoubtedly a variety of causes for the low levels of compliance but one fundamental reason is that it is more expensive to comply than are the costs of the likely consequences. In other words, most organizations appear to consider regulatory requirements as just another risk with low levels of impact that is just too expensive to mitigate. Given the generally lax enforcement and lack of regulatory teeth, security managers often find it difficult to mount a persuasive business case in terms of cost versus benefit for compliance.

> For half of the interviewed CIOs, IT Governance focuses on compliance, control and "operational IT," and is less about organisational aspects, architecture and decision structures. More mature organisations include the responsibility of the board and highest management layers, as well as the related decisionmaking structures, in their definitions. [2]

3.1 SECURITY GOVERNANCE AND REGULATION

Many of the regulations currently in effect either relate to or are directly addressed by governance, that is, by the framework of internal rules and structures that pro-

Table 3.1. Regulatory compliance levels

The percentage of U.S. organizations in compliance with:	
California Security Breach Notification Law	18%
Sarbanes–Oxley	35%
HIPAA, health care only	40%
Gramm–Leach–Bliley	14%
Other state or local privacy regulations	29%
Internationally, the averages were not a great deal better:	
Australian Privacy Legislation	50%
France CNIL	42%
U.K. Data Protection Act	31%
E.U. Privacy Directive	45%
Canadian Privacy Act	30%

vide guidance and control to the organization. They all deal with one or more of the following elements:

Disclosure
Transparency
Oversight
Record retention
Privacy
Attestation
Operational risk
Training

Most organizations will be subject to more than one of the current crop of regulations, which include but are not limited to:

- National Fire Protection Association (NFPA)
- Occupational Safety and Health Administration (OSHA)
- HIPAA
- European Union Data Protection Directive
- Copyright and patent laws for each country in which an organization performs business
- Office of the Comptroller (OCC), Circular 235 and Thrift Bulletin 30, Security Statutes (cover areas of U.S. computer fraud, abuse, and misappropriation of computerized assets), for example, the Federal Computer Security Act
- U.S. Federal Financial Institutions Examination Council (FFIEC) guidelines, which replaced previously issued Banking Circulars BC-177, BC-226, and so on
- COSO
- CoCo
- Cadbury
- King
- Foreign Corrupt Practices Act (FCPA)
- Vital records management statutes, specifications for the retention and disposition of corporate electronic and hardcopy records, for example, IRS Records Retention requirements in the United States.
- Gramm–Leach–Bliley
- Sarbanes–Oxley
- FISMA
- BASEL II
- California SB 1386 and other breach notice laws
- Patriot Act

REFERENCES

1. *The Global State of Information Security,* Price Waterhouse Coopers, 2006.
2. *IT Governance in Practice Advisory and Tax: Insight from Leading CIOs,* Price Waterhouse Coopers, 2006.

Chapter **4**

Roles and Responsibilities

Many organizations can markedly improve their security posture with greater emphasis and clarity regarding safety-related roles and responsibilities. It is axiomatic that those things for which no one is explicitly accountable are often ignored. In addition, a significant percentage of senior management are from the preinformation age and are not always well versed in security risks and benefits or their responsibilities in regard to information systems and security. Lacking this understanding, it is often difficult to get management focus and support for information security initiatives and the "tone at the top" may not be conducive to a security-oriented culture. A common manifestation of this problem is senior management not adhering to or supporting the security policies, thus setting a bad example and making it difficult to achieve compliance at lower levels of the organization.

It is widely recognized by security practitioners that without senior management buy-in and support for security activities, it can be difficult to achieve the level of security required to adequately address risks. Experience has shown that security simply cannot be driven upward from the middle when support from the top is lacking. Often, it appears that it takes a major compromise to provide a wake-up call and inspire the necessary focus. In some notable circumstances, this has led to the demise of the organization, such as occurred with Card Systems, a payments processor, or the Barings Bank as the result of losses incurred by the actions of a rogue currency trader.

In many cases, it may simply be a lack of management's awareness of the roles and responsibilities that need to be undertaken at various levels of the organization, and providing education may address the issue.

In general terms, the following section provides an overview of the roles and responsibilities required at different levels of the organization for effective information security governance.

4.1 THE BOARD OF DIRECTORS

Governing boards have a variety of responsibilities, including setting strategic direction, ensuring that risk is managed appropriately, ensuring that adequate resources are used responsibly, and performance measurement. It is a fundamental requirement that senior management protects the interests of the organization's stakeholders. Given the increasing criticality and near total dependence on information and the systems that process, transport, and store it, it is arguably an absolute requirement of due care to ensure that these assets are handled responsibly.

To quote Shirley M. Hufstedler, a former director of Hewlett-Packard, "The rising tide of cybercrime and threats to critical information assets mandate that boards of directors and senior executives are fully engaged at the governance level to ensure the security and integrity of those resources."

The National Association of Corporate Directors recognized these obligations by stating that essential security practices for directors include:

- Place information security on the board's agenda.
- Identify information security leaders, hold them accountable and ensure support for them.
- Ensure the effectiveness of the corporation's information security policy through review and approval.
- Assign information security to a key committee.

Table 4.1 provides an overview of some of the roles and responsibilities generally required of various levels in the organizational structure to achieve positive outcomes in each of the areas of information security governance.

Culture and "tone at the top" has been identified as one of the major contributors to organizational failures such as Enron, Worldcom, TJX, CardSystems, and others [2].

4.2 EXECUTIVE MANAGEMENT

Active support for security initiatives must come from management to maximize successful outcomes. Major security projects typically impact a broad swath of the organization and often encounter some resistance, which must be addressed with executive support. Without executive support, business unit leaders, who are often higher in the organizational structure than the security manager, may not have a great deal of incentive to support the program. Experience indicates that a lack of sufficient executive support is often due to the failure of security practitioners to develop a persuasive business case or a management not fully understanding the security issues and the benefits of effective security governance. Well-developed presentations in business terms coupled with executive awareness training may be helpful in garnering support for security activities. This must include specific busi-

Table 4.1. Basic information security responsibilities [1]

Management level	Strategic alignment	Risk management	Value delivery	Performance measurement	Resource management	Integration
Board of directors/ trustees	Set direction for a demonstrable alignment	Set direction for a risk management policy that applies to all activities and regulatory compliance	Set direction for reporting of security activity costs and value of information protected	Set direction for reporting of security effectiveness	Set direction for a policy of knowledge management and resource utilization	Set direction for a policy of assuring process integration
Senior executives	Institute processes to integrate security with business objectives	Ensure that roles and responsibilities include risk management in all activities, monitor regulatory compliance	Require business case studies of security initiatives and value of information protected	Require monitoring and metrics for reporting security activities	Ensure processes for knowledge capture and efficiency metrics	Provide oversight of all management process functions and plans for integration
Steering committee	Review and assist security strategy and integration efforts, ensure that business unit managers and process owners support integration	Identify emerging risks, promote business unit security practices, identify compliance issues	Review and advise adequacy of security initiatives to serve business functions and value delivered in terms of enabled services	Review and assure that security initiatives meet business objectives	Review processes for knowledge capture and dissemination	Identify critical business processes and management assurance providers, direct assurance integration efforts
Chief information security officer	Develop security strategy, oversee security program and initiatives, liaise with business unit managers and process owners for ongoing alignment	Ensure risk and business impact assessments, develop risk mitigation strategies, enforce policy and regulatory compliance	Monitor utilization and effectiveness of security resources and reputation and the delivery of trust	Develop and implement monitoring and metrics collection, analysis and reporting approaches, direct and monitor security activities	Develop methods for knowledge capture and dissemination, develop metrics for effectiveness and efficiency	Liaise with other management process functions, ensure gaps and overlaps are identified and addressed

ness linkages and supportable financial analysis indicating adequate cost/benefit ratios.

In many organizations, legal and regulatory requirements are the main drivers for efforts to achieve some level of compliance. Although these efforts can result in significant overall improvements in security, they are often approached as doing the minimum absolutely necessary with a "check the box" and move on perspective. As shown in Table 3.1, overall global compliance remains low and this is not encouraging for the security manager expecting management support for security initiatives. Nevertheless, given the continued spectacular security failures, increasing sophistication of attackers, and utter dependence on information systems, increasing compliance enforcement and evolving regulation is a certainty and will drive the necessity for management involvement.

4.3 SECURITY STEERING COMMITTEE

A properly constituted security steering committee can be of significant benefit to the achievement of effective information security governance. The participants are ideally senior representatives of the main operational and administrative functions in the organization. This should include business unit leaders as well as the HR, legal, finance, and marketing departments. The committee should have a charter and specific responsibilities as indicated in Table 4.1.

This approach can provide a number of benefits including a forum for identifying and prioritizing current and emerging risks, an invaluable channel for gathering organizational intelligence, as well as an avenue for disseminating important security-related information. The committee can be instrumental in gaining consensus to aid security program activities as well as serving as a forum for dispute resolution.

4.4 THE CISO

The responsibilities of information security managers varies widely as does their position in the organizational structure. The trend over the past few years has seen an increase in responsibilities and authority of security managers in many sectors, and many organizations have created the position of Chief Information Security Officer (CISO) or equivalent. There is also a trend toward integrating physical security with information security as there has been a belated realization that these issues are two sides of the same coin and information security absent physical security is not possible. This was highlighted recently in the Australian Customs Office when two individuals purporting to be service personnel assisted by data center personnel physically stole two highly sensitive, technically well-secured servers.

The elevation of security management to the executive has increased significantly in the past few years with the Aberdeen Group in its 2005 report titled *Best Practices in Security Governance* indicating that, "Some 40% of [2005] respondents reported that their companies employ a CISO or CSO, up from 31% in 2004." More recent studies indicate that this trend has subsequently stalled but this is likely to be a short-term situation. Financial institutions have generally blazed the trail for information security and the vast majority have executive-level information security management reporting to the COO or CEO, in part due to many central banks requiring it and in part due to the fact that security is a bank's stock in trade, and a loss of trust is not good for business.

Consider the approach undertaken by a major bank to align security with its business objectives. The organization was clear that it wanted to expand its operations online because the efficiencies gained from online banking had proven persuasive. It was also demonstrated by surveys and analysis undertaken by the CISO that customer trust was the single most important factor in driving growth and the factors that most affected that trust were primarily in the information security domain. In response to these findings, senior management implemented a plan whereby several hundred of the top managers have a substantial portion of their compensation tied to evaluation of their adherence to security requirements and compliance of their respective organizations.

Although roles and responsibilities will vary between organizations, a typical job description for an information security manager might take the following form.

General Purpose

The Information Security Manager serves as the process owner for all ongoing activities that serve to provide appropriate access to and protect the confidentiality and integrity of customer, employee, and organization information in compliance with policies and standards.

Position Responsibilities

- Serves as an internal information security consultant to the organization
- Documents security policies and procedures created by the Information Security Committee
- Provides direct training and oversight to all employees, affiliate marketing partners, alliances, or other third parties, ensuring proper information security clearance in accordance with established organizational information security policies and procedures
- Initiates, facilitates, and promotes activities to create information security awareness within the organization
- Performs information security risk assessments and serves as an internal auditor for security issues
- Implements information security policies and procedures for the organization
- Reviews all system-related security plans throughout the organization's network, acting as a liaison to Information Systems
- Monitors compliance with information security policies and procedures, referring problems to the appropriate department manager
- Coordinates the activities of the Information Security Committee

- Advises the organization and provides current information about information security technologies and related regulatory issues
- Monitors internal control systems to ensure that appropriate access levels are maintained
- Prepares a disaster recovery plan

This job description is typical for operational-level activities but does not address the strategic issues that need to increasingly be under the purview of the CISO or CSO for effective information security management aligned with and supporting the organization's business objectives.

Reporting

Approximately 35% of information security managers still report directly or indirectly to a CIO who is also responsible for IT and there are recent indications that this may be increasing. Although in many instances this is functionally adequate, it is structurally unsound. IT is about performance, whereas information security is about safety, which creates inevitable tensions and an inherent conflict of interest. Experience shows that the quest for greater IT performance at less cost is often made at the expense of security. The TJX debacle, with the subsequent loss of over 46 million credit records and at a cost to date of over $250 million U.S., exemplifies this situation, as revealed by court records.

REFERENCES

1. *Information Security Governance: Guidance for Boards of Directors and Executive Management,* 2nd Edition, ITGI, 2006.
2. *Enterprise Governance,* International Federation of Accountants, 2004.

Chapter 5

Strategic Metrics

Metrics are measurements from one or more points of reference and serve only one functional purpose—*to provide the basis for decisions.* Governance metrics are primarily navigational aids as opposed to performance measures and must provide information on strategic matters rather than management or operational issues, which are covered in Chapter 13. The purpose of governance metrics is to provide the information needed by senior management, including executive-level security managers, to make the decisions necessary for long-term guidance of a security program and assurance that it is operating in "the green."

There are generally two types of metrics: quantitative and qualitative or some combinations of the two. Though there will be general discussion of the types and nature of metrics in this book as related to the various levels of governance, in-depth analysis of metrics types, design, and deployment fills a separate volume by this author and is beyond the scope of this work. It is assumed that the reader has a general familiarity with the topic.

Security governance is not possible to any meaningful extent without metrics and the opposite is true as well. Governance requires defined objectives to know what we are to manage to. Metrics are essential to provide the feedback needed to determine if we are heading in the right direction as well as current location and proximity to the destination. Governance, in turn, provides the objectives that are the reference points for meaningful management metrics.

It is important to understand the distinction between strategic, management, and operational metrics. Though they are easy to obtain and abundant, most technical IT operational metrics are of little use in determining strategic direction or managing an information security program. This can be likened to the operation of an aircraft that has three types of basic instrumentation.

One is operational information regarding the machinery, such as oil pressure, fuel supply, temperature, and so forth, which is analogous to IT metrics.

The second is aircraft management information such as airspeed, attitude, heading, and altitude, which is needed to manage the aircraft properly but, ultimately, only relevant if the destination is known. Flying safely in circles is not likely to be very useful.

The third is navigational or strategic information including direction to the destination and position. All three types of information are necessary for proper operation and to meet the overall strategic objectives of the organization such as operating an airline.

Whether operating an airline, manufacturing widgets, or managing a security program, the issues are the same and the types of information required are as well. The majority of organizations nevertheless attempt to operate security using primarily operational information, which makes as much sense as flying aircraft without knowing position or destination, attitude or altitude.

Indeed, Andrew Jaquith of the Yankee Group expressed it well at the Metricon 1 metrics conference in 2006 during a keynote speech:

> Security is one of the few areas of management that does not possess a well-understood canon of techniques for measurement. In logistics, for example, metrics like "freight cost per mile" and "inventory warehouse turns" help operators understand how efficiently trucking fleets and warehouses run. In finance, "value at risk" techniques calculate the amount of money a firm could lose on a given day based on historical pricing volatilities. By contrast, in security . . . there is exactly nothing. No consensus on key indicators for security exists. [1]

In this chapter, we will explore the strategic, or "navigational" requirements for security and the metrics needed to "fly right". Management metrics are discussed in Chapter 13.

5.1 GOVERNANCE OBJECTIVES

At the risk of belaboring the point, we must define the objectives, or destination, of information security if we are to devise useful metrics needed to guide decisions necessary to get there. We have previously used the ISACA definition of governance:

> The set of responsibilities and practices exercised by the board and executive management with the goal of providing *strategic direction,* ensuring that *objectives are achieved,* ascertaining that *risks are managed* appropriately, and verifying that the enterprise's *resources are used responsibly.* [2]

Having a definition of governance does not tell us how to get it or measure it but it does provide the basis for setting the objectives needed to achieve it. However, for this definition to be of value for information security (or any other) governance, specific objectives and requirements must be defined for each of the four elements:

1. **Strategic direction** must be set for the organization generally and for an information security program specifically. Typically, this is the purview of the

governing board and executive management. As a practical matter, senior management may need the advice and guidance of the information security manager to define objectives for a security program but it still requires their acceptance and support of the program.

2. **Ensuring that objectives are achieved** obviously requires that those objectives are clearly defined and some method of monitoring and metrics be devised to provide that assurance.
3. **For risks to be managed appropriately** requires that senior management determines what the appropriate level of management of risk is (risk tolerance) and determines what monitoring and metrics will provide assurance that this is occurring.
4. **Verifying that resources are used responsibly** means that a determination must be made as to what measure will be used to gauge a suitable level of responsibility as well as a means of monitoring, and metrics must be developed to provide that assurance.

5.1.1 Strategic Direction

Since the fundamental purpose of information security is the protection of the organization and providing a reasonably predictable base for operations, its activities and goals must be aligned with the organization's long-term objectives if it is to provide value.

Strategic objectives may be formally documented and readily available or more obscure and only known to constituent business and operational units. Whatever the case, this information is needed to develop functional goals for an information security program, which in turn will frame the requirements for information security governance. Since, fundamentally, governance is the structure and system of rules for governing, it is essential that it be clear what it is designed to accomplish.

A typical example could be a bank with the long-term strategic objectives of eliminating most physical branches and converting the majority of its activities to online, internet-based operations. Clearly, the security requirements will change dramatically as this evolution occurs. Entirely new threats must be assessed, levels of acceptable risks must be determined, systems and requirements developed, and so forth. It would obviously be negligent to consider this business approach without an evaluation of risks, impacts, and mitigation approaches, as well as defining what the structure and rules of operation will be for the initiative. *The objectives must be defined for the systems to be designed and security aligned.*

5.1.2 Ensuring that Objectives are Achieved

Once objectives are defined and the strategic direction is set, ensuring that those objectives are achieved will require defining a strategy (Chapter 6) for implementation, milestones, and monitoring and metrics for governance feedback.

What metrics will provide the information needed for ensuring that objectives are achieved? First, there will be different phases in any initiative and there will be different information needed during those phases. A useful approach to considering the various phases is the system-development life-cycle method (SDLC). Depending on the source, there are some variations in the details of the SDLC approach but, generally, they are similar to the following:

- Feasibility
- Requirements
- Architecture and design
- Proof of concept
- Development
- Deployment
- Maintenance
- End-of-life decommissioning

Considering each of these steps, we can determine the types of strategic decisions that must be made and the information needed to make them, which, in turn, will define the metrics needed. At this level, many of the metrics will be roll-ups or synopses of various assessments and studies to determine whether an initiative is on time, will have the desired effects, requires additional resources, and so on. Generally, the information needed will be along the following lines:

- Risks—requires a risk assessment and analysis
- Effectiveness of measures to mitigate risk—assessment of control objectives and controls
- Value/benefits—requires studies of markets, competitions, and trends
- Impact of failure—business impact assessment from compromise or failure
- Total cost of ownership (TCO)—includes acquisition, deployment, maintenance, training, impact on productivity, and so on
- Return on investment—financial metrics such as IRR, NPV, and ROI

5.1.3. Risks Managed Appropriately

For risks to be managed appropriately will, of course, mean that what is "appropriate" is determined by senior management. It will also, by inference, mean at what cost. But unless it is determined what that means rather precisely, it will not be possible to manage it in any measurable way. Generally, the reference is to "risk tolerance" but that is still nebulous and must be defined to provide a point of reference.

One approach is to arrive at a management decision as to what monetary loss amount constitutes an "acceptable risk." If, for example, management determines that any single risk that cannot cause more than a $10,000 loss with a probable frequency (with a 95% certainty) of XX% or less annually is not worth the time and effort to mitigate, then there is a point of reference that can guide risk management efforts. Of course, assessment and analysis will be complex and arduous to make those determinations and will include risk and business-impact assessments and analysis as well as annual loss expectancy (ALE), return on security investment (ROSI), and, possibly, value at risk (VAR) computations (these and others are discussed in Chapter 13).

Another approach could be to perform the foregoing analysis first and then rank possible losses, probable frequency, maximum and probable single-loss events, and, perhaps, aggregation probability followed by total costs to mitigate impacts to various levels, along with methods of doing so. Management will then be in a position to decide what would be "appropriate" at what cost.

A third approach can be derived from business continuity planning (BCP) and developing recovery time objectives (RTOs). This will require, at a minimum, business impact assessment (BIA) and a risk assessment to determine risk level and probability. The determination of the criticality of a particular business process and understanding the impact of failure or compromise followed by the costs of unavailability over time will provide a basis for determining the recovery times needed to control impacts. Analysis of what will be required to recover the function within the necessary time will provide a method of determining the cost of doing so. Since shorter recovery times will usually be more expensive, evaluating the benefits versus the costs will reveal the optimal point at which the cost of losses equals the cost of recovery. Performing this exercise for all critical systems will provide a basis for determining optimal cost/benefit ratios of managing risk that will supportably be "appropriate."

There are two problems with this approach. One likely problem is whether management finds the costs acceptable, which experience indicates would not be typical. The other problem is that some aspects are inherently speculative and difficult to determine with certainty, such as the frequency and magnitude of the realization of potential risks.

5.1.4 Verifying that Resources are Used Responsibly

As with other aspects of the governance definition, "responsibly" must be clarified for any form of metrics or monitoring to be reasonably possible. It is a common term; most will have a reasonably good idea as to its meaning but it is difficult to define with any precision. General clarification can include various specifics such as:

- Using resources only for acceptable organizational purposes
- Achieving specified levels of utility
- Analysis of cost/benefit, showing acceptable levels
- Resource allocation based on risk reward analysis
- Achieving an anticipated return on investment (ROI)
- Realizing targeted productivity gains

There are undoubtedly many additional specifics that might be suitable, but these are typical and relatively straightforward to develop metrics for. Although many of the metrics suggested will be of significant value at the security management level, some roll-up or aggregation of this information will provide senior management with the information needed to determine if the program is on track and meeting objectives. The decisions that must be made at the strategic level will typically be whether to do nothing because objectives are being verifiably met, or whether something must be changed because they are not.

REFERENCES

1. Jaquith, A., Yankee Group, "Metrics are Nifty," Metricon, Metrics Conf., 2006, www.security metrics org/content/wiki.jsp?page=metricon1.0Keynote, 2006.
2. *Enterprise Governance,* International Federation of Accountants, 2004.

Chapter 6

Information Security Outcomes

There is a difference between objectives and outcomes, although they may ultimately be the same. Objectives set the targets for efforts; outcomes are the result. Obviously, we hope that achieving the objectives we set results in the outcomes we want.

Knowing the desired outcomes is important to defining the objectives but, to a large extent, the terms can be and are used interchangeably. In any event, initiating the implementation of information security governance as discussed in the preceding chapters is the first step in securing the outcomes we want from an information security program. Understanding and clarifying the outcomes or results we seek will provide both direction and guidance for the definition of specific objectives as well as a basis for determining whether those outcomes are being achieved.

6.1 DEFINING OUTCOMES

The outcomes set forth by the IT Governance Institute to define the expectations of security governance [1] are useful in determining the ultimate results we are seeking from a security program. They will help shape the objectives, set the goals, and clarify the destination, which is essential for developing a strategy for getting there. They include:

- Strategic alignment
- Risk management
- Business process assurance/convergence
- Value delivery
- Resource management
- Performance measurement

Having enumerated the desired outcomes of information security governance, it will still be necessary to determine the acceptable extent to which they must be achieved as well as what they mean in practical terms in order to develop specific objectives and a strategy to achieve them. Questions that need to be answered include:

- How much alignment with organizational objectives must security have? How do we define and measure it? Is alignment increasing or decreasing? How can it be measured?
- What levels must risk be managed to? What is the measure to determine when it has been achieved?
- What options are available to integrate assurance functions? How can the silo effect of safety efforts be countered? What would constitute an optimal level of integration?
- What is an adequate or optimal level of value delivery? How can it be improved? How can it be measured?
- How do we determine if resources are being used effectively and efficiently?
- What level of performance measurement will be sufficient to guide the security program and maintain an acceptable level of security?

Let us consider how these questions might be answered and what level of clarity we can bring to determining how these six outcomes can be defined to a level that can be managed and measurable. Though there are a number of possibilities, a common approach to create practical points of reference to gauge the extent to which these outcomes will be realized is to develop key goal indicators (KGIs). These goals need to be developed in conjunction with the organization's business and operational units to ensure relevance to their activities. Key goal indicators can be any form of metric, whether an actual numeric value such as the number of complaints in some period of time or periodic surveys of organizational sentiment regarding security. These indicators will provide useful feedback for security management for navigating the program and providing a general metric for the organization to monitor progress.

6.1.1 Strategic Alignment—Aligning Security Activities in Support of Organizational Objectives

How much alignment with organizational objectives must security have? How do we define and measure it? Is alignment increasing or decreasing? How can it be measured?

Strategic alignment is a straightforward concept with many implications and possible complexities. These can include very different perspectives on what constitutes optimal security support from a business perspective as opposed to the viewpoint of security. However, any chance of achieving a reasonable balance will require security managers to understand the business objectives and concerns and find the least disruptive approach to maintaining adequate safety. With these caveats, what measurement of outcomes can be defined that will provide assurance of achievement?

A working definition of the relative degree of alignment can be the extent to which any security activity furthers or hinders any particular organizational activity. Since the objective will be to provide maximum cost-effective support for the activities of the organization, consistent with maintaining adequate security, we will need to know what that means in practical terms. It will be important to understand what maximum cost-effective support will look like and what it must accomplish. Certainly, an acceptable level of safety, predictability, reliability, and integrity for business operations is likely to include some of the attributes that must be considered.

These goals will, by necessity, be primarily qualitative since there are no hard measures of the "degrees of alignment."

It must be kept in mind that to a significant extent there may be a need to manage perceptions and to overcome a predisposition to see security as a "business preventer" rather than an enabler. These perceptions will, to an extent, be a function of the department's position in the organization hierarchy. Operations are typically more focused on performance than on safety. Senior management wants performance, minimal impacts, and predictability. Lower levels of the organization may see security as simply a nuisance to be circumvented when possible.

Some possible KGIs for strategic alignment could include:

- **Lines of business have defined security requirements and control objectives.** Security activities and control requirement are likely to be different for various organizational activities. The KGI could be identified security activities and control objectives for each operational group.
- **Defined business requirements drive security initiatives.** For information security to be aligned with business, it must be driven by business requirements. The goal indicator would be a substantive business case for all significant security initiatives including risk, impact, and cost/benefit analysis.
- **Security activities do not materially hinder business.** Whether this is true or the general organizational perception is not important. A growing sentiment that security hinders business will have a negative impact on the ability to be effective. The KGI could be a periodic survey to see if there is a developing sentiment that security is interfering with business activities. Another KGI for alignment could be the number of policy exceptions indicating that security standards are posing an operational problem.
- **Security program enables certain business activities.** Recognition that certain business activities could not be undertaken without appropriate security will benefit the view of alignment. Periodic surveys could be used for this assessment to gauge the trends as well.
- **Security activities provide predictable operations.** Predictability is good for business. A KGI could be the extent that incidents and impacts fall within an anticipated and acceptable range.

- **Security resources are allocated in proportion to business criticality.** Alignment with business objectives suggests that the most critical business operations receive the greatest protection. The KGI could be demonstrable proportionality of resource allocation and asset criticality and/or sensitivity.

6.1.2 Risk Management—Executing Appropriate Measures to Manage Risks and Potential Impacts to an Acceptable Level

What levels must risk be managed to? What is the measure used to determine when it has been achieved?

Management of risk presents a number of challenges for the information security manager. For one thing, there is typically not enough information to determine "acceptable levels" with any degree of precision, and, for another, actually determining the degree of risk despite years of study and development is more art than science. Certainly, at a statistical level we can determine the probable degree of risk, provided there is a large enough sample. Given a large population of individuals, we know how long the average individual will live and how many traffic accidents they will have. Unfortunately, statistical averages are of little use in determining individual situations.

The responsibilities and objectives of risk management are well described in the ISACA 2008 Certified Information Security Manager Review Manual [2]:

> The objective of this job practice area (risk management) is to ensure that the information security manager understands the importance of risk management as a tool for meeting business needs and developing a security management program to support these needs. While information security governance defines the links between business goals and objectives and the security program, security risk management defines the extent of protection that is prudent based on business requirements, objectives, and priorities.
>
> The objective of risk management is to identify, quantify and manage information security-related risks to achieve business objectives through a number of tasks utilizing the information security manager's knowledge of key risk management techniques. Since information security is one component of enterprise risk management, the techniques, methods and metrics used to define information security risks may need to be viewed within the larger context of organizational risk.

Managing risk effectively is complex and this complexity is compounded by the fact that risk management responsibilities are usually split between a number of organizational entities, with the consequence that the biggest risk may well be a lack of continuity and integration between these efforts. The fact that all parts of any organization are required to operate in some fashion related to managing risk further complicates the problem. Though many of these risk management concerns may be the responsibility of an organizational risk manager, most will also have a direct impact on information security. This includes most elements of physical security, including how every user of information systems behaves, how physical information is handled, how laptops and other portable devices storing information are managed, and how access to facilities is controlled, to mention a few.

Good metrics directly informative of risk are nonexistent, at least in the area of information security. Although we can get more-or-less direct metrics on technical vulnerabilities, those on most other components of risk, including procedural vulnerabilities, threats, frequency, probability, and magnitude, will not be as simple to obtain. Risk assessments are the primary approach to ascertaining risks but are highly speculative; they are only a snapshot in time and just a form of monitoring. They are also subjective and imprecise, which results in the likelihood that some risks will be overestimated and excessive precautions taken, or that the underestimation of risks will ultimately result in unfortunate consequences.

Although the increasingly sophisticated approaches such as value-at-risk (VAR) computations and other complex analysis methods offer the promise of providing better risk management metrics, most are not ready for general implementation and their utility remains to be demonstrated. In most situations, technical vulnerability scans, security reviews, audits, and monitoring are typically the only viable options.

Against this backdrop of high hurdles, it is nevertheless an imperative that risks must be managed and they are often managed quite successfully. It is probably helpful to be lucky as well.

The decisions that must be made by an information security manager about managing risk are numerous, complex, and generally lacking information of sufficient clarity and precision for any degree of certainty. These decisions are often guided by intuition and experience. They are, of course, also sometimes wrong.

The range of decisions that are typically most important are the type and level of protection that should be afforded various information-related assets and whether the protection provided is in fact adequate. The whole notion of layered security shown in Table 6.1 is to compensate for the inherent uncertainty in the entire risk assessment and management process.

Though well known to practitioners, it may be useful to dissect the kind of information needed to make rational decisions about managing risks, including:

- Criticality of assets
- Sensitivity of assets
- The nature and magnitude of impact if assets are compromised
- The extent and types of vulnerabilities and conditions that may change them
- The extent and nature of viable and emerging threats
- The probability or likelihood of compromise
- Strategic initiatives and plans
- Acceptable levels of risk and impact
- The possibility of risk aggregation or cascading

Key goal indicators from a governance perspective can be used to indicate whether we are heading in the right direction to appropriately manage risk. Some possibilities to consider can include:

Table 6.1. Layered security

Defenses against system compromise	Policies, standards, procedures, and technology
Prevention	Authentication
	Authorization
	Encryption
	Firewalls
	Data labeling/handling/retention
	Management
	Physical security
	Intrusion prevention
	Virus scanning
	Personnel security
Containment	Awareness and training
	Authorization
	Data privacy
	Firewalls/security domains
	Network segmentation
	Physical security
Detection/notification	Monitoring
	Measurements/metrics
	Auditing/logging
	Honeypots
	Intrusion detection
	Virus detection
Reaction	Incident response
	Policies/procedure change
	Additional security mechanisms
	New/better controls
Evidence collection/event tracking	Auditing/logging
	Management/monitoring
	Nonrepudiation
	Forensics
Recovery/restoration	Backups/restoration
	Failover/remote sites
	Business continuity/disaster recovery planning

Source: Brotby, Krag; "Xerox BASIA Architecture," 1996.

- **Complete Periodic Risk Assessment.** Despite decades of promoting the necessity for risk assessments, the astonishing fact is that roughly half of all organizations have not done so, including some major financial institutions known to the author. Risks are not likely to be well managed unless they are known. A KGI would be the performance of periodic risk assessments.
- **Business Impact Assessment.** An even greater percentage of organizations do not engage in business impact assessments and analysis. It is unlikely that risk management efforts will be effectively prioritized or appropriate re-

sources allocated without understanding the potential impacts of compromise. A KGI would be periodic impact assessments of all critical systems.

- **Business Continuity Planning/Disaster Recovery (BCP/DR).** It is certain that organizations that do not do risk assessments and BIAs have not developed BCP/DRP of any consequence. If they do exist, one KGI would be tested BCP/DR and another would be maintenance of consistent updates to the plans.
- **Defined Risk Appetite.** To manage risk appropriately, the level of acceptable risk must be decided in order to know what to manage risk to. A KGI would be defining risk tolerance in terms of maximum acceptable impacts or losses and, perhaps, an acceptable ratio of probable losses to costs of mitigation.
- **Asset Classification.** Risk management cannot be apportioned appropriately without determining the level of sensitivity and criticality of information assets. A KGI would be documented assignment of asset ownership and classification as to sensitivity and criticality.

Other possible KGI's can include such things as:

- An overall security strategy and program for achieving acceptable levels of risk
- Defined mitigation objectives for identified significant risks
- Processes for management or reduction of adverse impacts
- Systematic, continuous risk management processes
- Trends of periodic risk assessment, indicating progress toward defined goals
- Trends in impacts
- Analysis of collective impact of aggregated risk
- Recognition of the potential for cascading impacts

6.1.3 Business Process Assurance/Convergence—Integrating All Relevant Assurance Processes to Improve Overall Security and Efficiency

What options are available to integrate assurance functions? How can the silo effect of safety efforts be countered? What would constitute an optimal level of integration?

An increasingly important area of concern, driven in large part by the increasing tendency to segment security into separate but related functions, focuses on the integration of an organization's assurance processes related to safety or security. Clearly, most or all of these activities will have relevance for security, both operationally and from a management perspective. In large organizations, these functions may include:

- Risk management
- Audit

- Legal
- HR
- Insurance
- Training
- Privacy
- Compliance
- Quality assurance
- Facilities/physical security
- Project management
- Help desk
- DR/BCP
- Forensics
- Architecture

The division of security-related functions into a myriad of separate activities with differing scope, mandates, and reporting structures is generally arbitrary. It typically has no relationship to the task of efficiently and effectively providing overarching assurance of organizational safety and risk management. Results often include overlapping or contradictory security initiatives, which waste resources, or major gaps that can lead to serious security compromises. An illustrative example is a recent case in which two individuals posing as repair technicians physically stole two electronically well-secured, highly sensitive database servers from an Australian customs office. Another involved a technically secure network used to process fraudulent orders. In both cases, the lack of integration of assurance management processes left large gaps, resulting in serious impacts. Part of the problem management guru Peter Drucker explained is that "every layer of management doubles the noise and cuts the useful information in half." This notion is even more true when arbitrary divisions of continuous processes are fragmented.

It is this fragmentation and the realization that it markedly degrades the ability to provide overall assurance of safety that has led to the notion of "convergence" of physical, IT, and information security that is the focus of a group comprised of ASIS, ISACA, and ISSA. It has, on a national level, been the driver for the creation of the U.S. Department of Homeland Security, with the mandate of integrating a number of previously independent security and safety related departments. This, surprisingly, is a new concept not yet widely recognized and supported, which gives rise to a number of questions when considering this issue:

- What level of integration of assurance processes should the information security manager strive to achieve?
- What would be a minimal level that might be acceptable? The case can be made that total integration is a worthy objective, however unrealistic.
- What is the business case for attempting to achieve this integration?

- What then are approaches to achieving an acceptable level of functional (if not actual) integration of the activities that provide assurance of security or safety?
- What measures can be used to gauge the level of integration on an ongoing basis?
- What metrics can be devised to determine the optimal level?

It is beyond the scope of this book to address all of these issues but if management has determined that this is a worthwhile effort, there are metrics and measures that may be useful indicators of the level of functional or operational integration. These can, in turn, be used to guide decisions about whether greater impetus is needed to achieve an appropriate level of integration of assurance functions.

The decision regarding whether there is a functional degree of integration between assurance functions can be assessed by whether there are any drivers for it, such as explicit policy directives or mandates in the assignment of roles and responsibilities, and whether there is a level of prescribed communication between these functions, to the extent that there is roll-up of reporting to a common point in the organizational structure. In some organizations, many or all of these functions may be represented in an executive committee, in which case raising the issue at that level may provide an approach to providing a level of assurance integration.

As with other aspects of security (and all) management, it will be necessary to define objectives for the desired level of assurance integration based on the desired outcomes. It will then be necessary to determine if that level has been achieved and establish a process to monitor and maintain ongoing integration.

KGIs for assurance process integration could include:

- **Incidents, or a lack of them, traceable to a lack of integration.** The Australian Customs office example, in which two people appearing to be service personnel simply entered a supposedly high-security area and walked out with several highly sensitive servers, illustrates the point. Most organizations would hopefully consider something less dramatic as a suitable indicator of the need to coordinate and integrate these functions.
- **The number of management levels before assurance processes fall under the same organizational position.** It is axiomatic that the more reporting levels for the various assurance functions before coalescing to a single "authority," the greater is the likely lack of integration between their functions. A KGI could be fewer reporting lines and levels before reporting to a single organizational position.
- **Inconsistencies or contradictions in the objectives, policies, and standards applied to various assurance functions.** Various assurance functions should have common strategic objectives to avoid working at cross purposes. A KGI would be policies or documented references to functional interfaces between assurance processes in roles and responsibilities.
- **An absence of communications between assurance providers.** Better communications between assurance providers as well as collective participation

in a steering or risk committee can improve integration of activities. A KGI might be regular formal review meetings to promote collective understanding of risks and various departmental objectives, activities, and processes.

6.1.4 Value Delivery—Optimizing Investments in Support of Organizational Objectives

What is an adequate or optimal level of value delivery? How can it be improved? How can it be measured? How can the objectives of providing an optimal investment strategy for security be defined? What levels can be determined that will be considered suitable outcomes?

There are many possible elements to consider. The first step may be to develop measures or metrics to determine whether existing security controls are adequate and cost-effective. Adequacy must be determined in light of the organization's risk tolerance, as previously discussed, perhaps in relation to actual incidents and impacts. If impacts have been excessive, it is an obvious indication of inadequate controls. The more difficult situation is when there may have been no significant incidents but a risk assessment indicates an unacceptably high level of risk and the potential for serious consequences. In this event, the security manager is likely not to find much support for tightening controls and increasing security expenditures.

In any event, there are a number of ways that resource allocations might be improved. For example, control analysis may show that some costly controls are not particularly effective or may not address control objectives, and can be replaced with more cost-effective controls aligned with control objectives.

As in any other activity, the degree of standardization can be a useful indicator of value insofar as multiple solutions to the same problems will be more costly. For example, standardized access controls will be easier to manage and monitor than ones that are different in various parts of the organization. From a security perspective, homogenous controls, though easier to manage, will also add a dimension of risk in that a common vulnerability will aggregate risk. The benefits of standardization will, therefore, to some extent be offset by the need for greater robustness to offset the increased risk associated with a common failure mode.

It may be warranted to attempt to determine financial return on investment of new or existing security investments to provide the information needed to prioritize efforts for greatest cost-effectiveness. Determination of financial returns on security activities are inevitably speculative to some extent. Some view security more as an insurance policy and maintain that trying to calculate a return is not productive. Yet it must be recognized that virtually all organizational activities are guided by some form of cost–benefit analysis and, ultimately, security activities must be as well.

Some types of security investments readily lend themselves to financial analysis, whereas others pose more of a challenge. The cost savings for automating certain activities can be easily calculated. The financial benefits of preventing unpredictable and uncertain events will be more difficult to do with any supportable accuracy. It may be the problem of proving a negative but there have been efforts that may be helpful. ROSI proposes to assess return on security investments by the amount of the reduction of losses in relationship to the investment. ROSI considers the risk exposure or annual loss expectancy (ALE), which is the expected losses times the frequency of occurrence, typically a year, times the percentage of risk mitigation or reduction, minus the solution cost, divided by the solution cost:

ROSI = [(Risk Exposure × % Risk Mitigated) – Solution Cost]/Solution Cost

The weakness of the approach is the high degree of guesswork involved in determining the risk exposure as well as the extent to which a particular solution will reduce either frequency of occurrence of magnitude of impact.

The reduction of losses will often be speculative but in some cases supportable or even accurate. Historical data may show fairly consistent levels of losses over time that a specific course of action can reduce to a definable extent. The decisions will again rest on the most cost-effective means of achieving security objectives.

KGIs for value delivery can include:

- **Security activities are designed to achieve specific strategic objectives.** Security investments and activities are often guided by reactive tactical goals rather than strategic objectives. A KGI could be defined business objectives for investments and activities as well as cost–benefit analysis.
- **The cost of security being proportional to the value of assets.** Security costs and resource allocation as compared to asset criticality and sensitivity can be a KGI for suitable proportionality. A reasonable degree of proportionality suggests that a rational, consistent approach is in place for security information assets.
- **Security resources are allocated by degree of assessed risk and potential impact.** Studies show little if any relationship between security risks and losses and the allocation of security resources in most organizations. A KGI would be resource allocation based on risk or better, potential impact.
- **Controls are based on defined control objectives and are fully utilized.** A KGI would be defined control objectives and controls demonstrably meeting those objectives.
- **An adequate and appropriate number of controls to achieve acceptable risk and impact levels.** Controls usually grow organically over time, typically as a reaction to a specific event or incident as opposed to any strategic plan or objective. A KGI for the minimum number of controls that meet defined control objectives would be useful.
- **Control effectiveness that is determined by periodic testing.** Effectiveness of controls often degrade over time and assurance requires testing. A KGI could be the level of tested control effectiveness.

Other possible KGI's could include:

- Policies in place that require all controls to be periodically reevaluated for cost, compliance, reliabilitiy, and effectiveness

- Controls utilization—controls that are rarely used are not likely to be cost-effective
- The number of controls to achieve acceptable risk and impact levels—fewer effective controls can be expected to be more cost-effective than more less-effective controls

6.1.5 Resource Management—Using Organizational Resources Efficiently and Effectively

How do we determine if resources are being used effectively and efficiently? What measures are available for the information security manager to determine effective and efficient use of security resources? Considering the stated requirement, effectiveness can only be evaluated against defined objectives. Efficiency would require that those objectives were achieved at the lowest cost.

The set of measures of effectiveness relate to the prior notion of value with regard to cost-effectiveness. However, it is more related to operational concerns, that is, the ongoing utility and operational costs of managing security. Once again, absent objectives, this will be difficult to measure.

Some of the decisions that the security manager may need to make are based on the following issues:

- What would constitute effective and efficient resource management?
- What measures would be useful?
- Can existing resources be used to greater effect?
- What processes might be put in place that ensure effective and efficient use of resources?
- How would those processes be monitored or measured?

KGI's can include:

- **Benchmarking.** If better security (fewer losses related to security failures) is achieved at lower cost, the argument could be made that it is the result of better resource management. There could, of course, be other factors but they would certainly be supportive of this assertion. The KGI would be benchmarking against comparable organizations.
- **Resource utilization.** Measures of staff productivity could be used as an indicator of resource management. The KGI would be improved utilization metrics.
- **Number of Controls.** Reductions in the number of controls required to meet objectives would be a good indicator and an ongoing metric. The KGI would be the number and operating costs of controls required to meet control objectives.
- **Standardized processes.** Standardized processes will be more efficient than custom solutions. The KGI could be reducing the number of nonstandard solutions used.

6.1.6 Performance Measurement—Monitoring and Reporting on Security Processes to Ensure that Objectives are Achieved

What level of performance measurement will be sufficient to guide the security program and maintain an acceptable level of security? What are the requirements for performance measurement and monitoring? Assuming that it is apparent that it will be difficult to manage what is not being measured, decisions must be made about the state of performance measurement itself, based on information regarding some of the following questions:

- Are there adequate performance measures in place for strategic, tactical, and operational elements?
- Are all key controls monitored in some fashion that will indicate effectiveness?
- Are there measures of key control reliability?
- Will there be a clear and timely indication of control failure?
- Is the management process itself measured and monitored?

KGI's can include:

- **The time it takes to detect and report security-related incidents.** Incident response is typically more effective the sooner it responds before an incident becomes a problem. KGIs will be the times it takes to detect, respond to, and resolve incidents.
- **The number and frequency of subsequently discovered unreported incidents.** Undetected incidents and compromises poses a considerable risk as evidenced by frequent headlines such as TJX, which did not discover the theft of over 46 million credit card records for, reportedly, over a year. The KGI would be similar to the foregoing one in terms of the acceptable time to detect incidents or incidents subsequently uncovered by external auditors or others.
- **The ability to determine the effectiveness, efficiency, accuracy, and reliability of metrics.** Metrics are only useful to the extent that they are accurate, reliable, and cost-effective. The KGI is metrics on management and operational metrics in terms of validating accuracy, reliability, and cost-effectiveness.
- **Clear indications that security objectives are being met.** All defined objectives need to have predetermined measures of success. The KGIs are those measures agreed to indicate success of any particular activity.
- **Threat and vulnerability management.** Knowledge and monitoring of impending threats and vulnerabilities is important to successful outcomes. The KGI would be processes to track and monitor emerging threats and vulnerabilities and assess potential exposure and impacts.
- **Consistency or effectiveness of log review practices.** Log reviews are a useful monitoring and detection practice frequently not performed. The KGI

would be the extent to which logs are systematically reviewed or, perhaps, the number of significant events uncovered by log reviews.

REFERENCES

1. Brotby, Krag, *Information Security Governance, A Guide for Boards of Directors and Executive Management,* 2nd ed. ITGI, 2006.
2. Information Security and Control Association, *2008 CISM Review Manual.*

Chapter 7

Security Governance Objectives

We defined six outcomes of information security governance in Chapter 6 and suggested ways to gauge whether a security program is meeting expectations using a series of Key Goal Indicators. As previously stated, objectives can be considered the targets of our activities designed to result in the desired outcomes. These objectives will require much greater detail to allow us to determine the plans and methods needed to achieve the outcomes. Just as the desired outcome of creating a bridge over a river that will carry four lanes of traffic will say nothing about how it is to be designed or built, defining the outcomes of information security will say nothing of the processes, methods, or plans necessary to achieve them. We will need far more exacting descriptions of the structure and operation to set achievable objectives.

Since it is not possible to quantify an overall information security program to any significant extent, a useful approach is to describe a "desired state," essentially a snapshot at some future point of the essential elements, aspects, and operations of the program in terms of characteristics and attributes. Each of the desired outcomes previously discussed will need to be considered in detailed terms to describe "desired state" which sets the parameters for defining objectives, namely:

- *Strategic alignment*—aligning security activities with business strategy to support organizational objectives
- *Risk management*—executing appropriate measures to manage risks and potential impacts to an acceptable level
- *Business process assurance/convergence*—integrating all relevant assurance processes to maximize the effectiveness and efficiency of security activities.
- *Value delivery*—optimizing investments in support of business objectives
- *Resource management*—using organizational resources efficiently and effectively

- *Performance measurement*—monitoring and reporting on security processes to ensure that business objectives are achieved

Here, we will consider methods to describe in some level of detail what objectives will be required to achieve those outcomes. These can include elements such as control objectives, acceptable risk and impact levels, performance levels, and operational relationships. These are often described as people, processes, and technologies.

In our quest for defining objectives for information security, there are a number of approaches available to the security manager to define this "desired state" of security. Some are quite complex, such as the development of a comprehensive security architecture; other approaches are simpler. The decision about how best to proceed will be a function of the organization, its culture, complexity, risk appetite, resources, constraints, maturity, and perhaps other factors. Whether a particular methodology is used or an internal method is developed is not important. What is essential is that the objectives are defined with sufficient clarity so that the direction is clear, proximity to the destination can be determined, and the desired outcomes achieved.

In the next sections, we will look at five approaches to defining information security objectives including:

1. Security architecture
2. CobiT
3. Capability Maturity Model
4. ISO/IEC 27001, 27002
5. National Cyber Security Summit Task Force Corporate Governance Framework

7.1 SECURITY ARCHITECTURE

The most promising (and probably the most arduous) approach to describing the desired state of security that is nevertheless gathering adherents and increasingly being mandated is the development of security architecture. Management in some industries and government organizations are starting to realize that the sheer complexity of attempting to provide assurance of organizational safety and the reliable operation of utterly critical technologies involving hundreds or thousands of moving parts cannot be done in an ad hoc, reactive, unplanned manner. There is also belatedly a growing understanding that the myriad of assurance functions related to "security" are rapidly going in the direction of ever narrowing specialization with their own jargon, standards, methods, and associations, with little that serves to integrate these activities into a collective effort.

High-level security architecture (i.e., contextual and conceptual) is one activity that can effectively serve as the framework to address this issue. There is, however, still not a general understanding of what security architecture is or why an organization might want one. The following analogy from the Meta Group provides a useful explanation.

> In many respects, the information security architecture is analogous to the architecture associated with buildings. It begins as a concept, a set of design objectives that must be met (e.g., the function it will serve; whether it will be a hospital, a school, etc.). It then progresses to a model, a rough approximation of the vision forged from raw materials (read: services). This is followed by the preparation of detailed blueprints, or tools that will be used to transform the vision/model into a real and finished product. Finally there is the building itself, the realization, or output, of the prior stages. [1]

Given the increasing complexity and size of organizations, growing cyber risks and losses, increasing regulatory pressures, and ever more problematic security administration, it is suprising that to date, security architecture generally has had little impact on enterprise security efforts:

> Indeed, without one, evidence suggests that enterprises will default to a haphazard, reactive, tactical approach to constructing a secure environment, regrettably wasting resources and introducing more vulnerabilities as they proceed to fix others. [1]

The failure of organizations to embrace the notion of security architecture appears to have several identifiable causes. One is that such projects are expensive and time-consuming and there is little or no understanding or appreciation at most organizational levels for their necessity or potential benefits. Another reason is that security continues to be seen as primarily a technical compliance issue rather than a strategic business activity. Another factor may also be that there is not an abundance of competent architects who have sufficiently broad and deep experience to address the wide range of issues necessary to ensure a reasonable degree of success, nor, until recently, the resources and training to develop them.

The effect of this lack of "architecture" over time has been to have functionally less security, less integration, and increased vulnerability across the enterprise at the same time that technical security has seen significant improvement. This lack of integration contributes to the increasing difficulty in managing enterprise security efforts effectively.

Contemporary notions of security architecture, the rationale for them, and the benefits they afford are best described by the SABSA Institute [2] as described in the following subsections.

7.1.1 Managing Complexity

One of the key functions of "architecture" as a tool of modern business is to provide a framework within which complexity can be managed successfully. As the size and complexity of a project grows, many designers and design influences must all work as a team to create something that has the appearance of being designed by a single "design authority." As the complexity of the business environment grows, many business processes and support functions must all integrate seamlessly to provide effective services and management to the business, its customers and its partners. Architecture provides a means to manage that complexity.

7.1.2 Providing a Framework and Road Map

Architecture also acts as a road map for a collection of smaller projects and services that must be integrated into a single homogenous whole. It provides a framework within which many members of large design, delivery, and support teams can work harmoniously, and toward which tactical projects can be migrated.

7.1.3 Simplicity and Clarity through Layering and Modularization

In the same way that conventional architecture defines the rules and standards for the design and construction of buildings, information systems architecture addresses these same issues for the design and construction of computers, communications networks, and the distributed business systems that are required for the delivery of business services. As with the conventional architecture of buildings, towns, and cities, information systems architecture must take account of:

- The goals that are to be achieved through the systems
- The environment in which the systems will be built and used
- The technical capabilities of the people to construct and operate the systems and their component subsystems

7.1.4 Business Focus Beyond the Technical Domain

Information systems architecture is concerned with much more than mere technical factors. It is concerned with what the enterprise wants to achieve and with the environmental factors that will influence those achievements. In some organizations, this broad view of information systems architecture is not well understood. Technical factors are often the main ones that influence the architecture, and under these conditions the architecture can fail to deliver what the business expects and needs.

7.1.5 Objectives of Information Security Architectures

The underlying notion for all architecture is that the objectives of complex systems must be comprehensively defined; precise specifications developed; their structures engineered and tested for form, fit and function; and their performance monitored and measured in terms of the original design objectives and specifications. Few would argue that today's information systems are complex and comprised of an enormous number of moving parts. The fact that they operate at all and often fairly well is not an argument against the increasingly desperate need to bring architectural discipline and engineering rigor to the deployment of these systems.

Although a number of approaches to architecture for components of information systems exist (e.g., architectures for data and databases, servers, technical infrastructures, identity management, etc.), little exists for the overall comprehensive enterprise security machinery, its management, and its relationship to business objectives. An obviously absurd analogy would be if there were separate, unintegrated architectures and designs for aircraft wings, engines, navigational equipment, passenger seats, and so on, but no architecture and design for the complete aircraft and how the various components fit together. The result would be unlikely to function well if at all, and few would be inclined to trust it with their lives. Admittedly, information systems are designed to operate in a far more loosely coupled fashion, but the point is nevertheless relevant.

The Sherwood Applied Business Security Architecture (SABSA), has been developed during the past decade to address this issue and offers useful insights and approaches to dealing with many current design, management, implementation, and monitoring issues. The approach is a framework that is compatible with and can utilize CobiT as well as ITIL and ISO/IEC 27001. Although the approach in totality may be more sophisticated and complex than many organizations are prepared to deal with, with the increasing reliance on increasingly complex systems coupled with the growing problems of manageability and security, they will at some juncture have little choice but to "get organized."

For the security manager, the approach may be helpful in developing longer-term objectives or suggesting approaches to addressing current issues. Some organizations have adopted elements of the architectural approach piecemeal, with the long-term objective of full implementation over time. A summary description of the SABSA approach follows.

The SABSA Model comprises six layers, the summary of which is in Table 7.1. It follows closely the work done by John A. Zachman in developing a model for enterprise architecture, although it has been adapted somewhat to a security view of the world. Each layer represents the view of a different player in the process of specifying, designing, constructing, and using business systems.

There is another configuration of these six layers, which is perhaps more helpful, shown in Figure 7.1. In this diagram, the "operational security architecture" has been placed vertically across the other five layers. This is because operational secu-

Table 7.1. The SABSA model for security architecture development

The Business View	Contextual Security Architecture
The Architect's View	Conceptual Security Architecture
The Designer's View	Logical Security Architecture
The Builder's View	Physical Security Architecture
The Tradesman's View	Component Security Architecture
The Facilities Manager's View	Operational Security Architecture

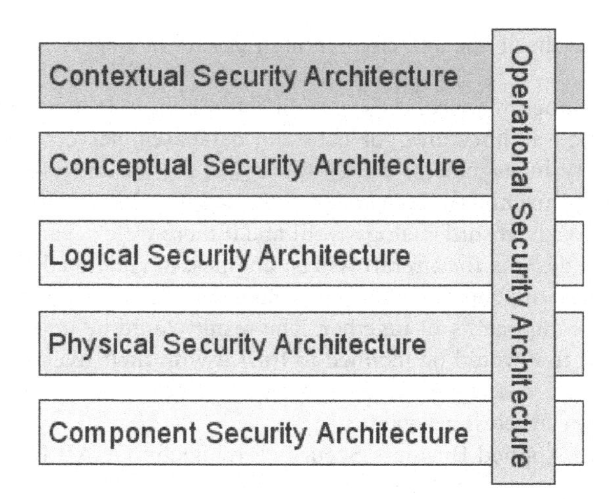

Figure 7.1. The SABSA Model.

rity issues arise at each and every one of the other five layers. Operational security has a meaning in the context of each of the other layers.

For detailed analysis of each of the six layers, the SABSA Matrix also uses the same six questions that are used in the Zachman framework: What, Why, When, How, Where, and Who? For each horizontal layer, there is a vertical analysis as follows:

1. What are you trying to do at this layer? The assets to be protected by your security architecture.
2. Why are you doing it? The motivation for wanting to apply security, expressed in the terms of this layer.
3. How are you trying to do it? The functions needed to achieve security at this layer.
4. Who is involved? The people and organisational aspects of security at this layer.
5. Where are you doing it? The locations where you apply your security, relevant to this layer.
6. When are you doing it? The time-related aspects of security relevant to this layer.

These six vertical architectural elements are now summarized for all six horizontal layers. This gives a 6 × 6 matrix of cells (Figure 7.2), which represents the whole model for the enterprise security architecture. If each issue raised by these cells can be addressed, there will be a high level of confidence that the security architecture is complete. The process of developing an enterprise security architecture is a process of populating all thirty-six cells.

	Assets (What)	Motivation (Why)	Process (How)	People (Who)	Location (Where)	Time (When)
Contextual	The Business	Business Risk Model	Business Process Model	Business Organization Relationships	Business Geography	Business Time Dependencies
Conceptual	Business Attributes Profile	Control Objectives	Security Strategies and Architectural Layering	Security Entity Model and Trust Framework	Security Domain Model	Security-Related Lifetimes and Deadlines
Logical	Business Information Model	Security Policies	Security Services	Entity Schema and Privilege Profiles	Security Domain Definitions and Associations	Security Processing Cycle
Physical	Business Data Model	Security Rules, Practices, and Procedures	Security Mechanisms	Users, Applications, and the User Interface	Platform and Network Infrastructure	Control Structure Execution
Component	Detailed Data Structures	Security Standards	Security Products and Tools	Identities, Functions, Actions, and ACLs	Processes, Nodes, Addresses, and Protocols	Security Step Timing and Sequencing
Operational	Assurance of Operational Continuity	Operational Risk Management	Security Service Management and Support	Application and User Management and Support	Security of Sites, Networks, and Platforms	Security Operations Schedule

Figure 7.2. The SABSA Matrix for security architecture development.

7.1.6 SABSA Framework for Security Service Management

The area of security service management, administration, and operations is addressed through the SABSA operational architecture layer. This layer of the framework is applied vertically across all of the other five, providing flexibility to ensure seamless and holistic integration with the standards and operational frameworks selected (Figure 7.3). This ensures information security compliance with frameworks such as ITIL, BS15000/AS8018, ISO 27002, and CobIT. SABSA provides the road map to determine how the requirements of these standards can be applied in individual business contexts.

7.1.7 SABSA Development Process

The SABSA Model provides the basis for an architecture development process, since it is clear that through understanding the business requirements, the architect can create the initial vision. This is used by the designers to create the detailed design, which, in turn, is used by the builder to construct the systems, with components of various sorts provided by specialists. Finally, the facilities manager (in this context that might be an IT manager, systems administrator, etc.) operates the finished system, but unless the earlier phases take account of the operational needs, this phase in the lifetime of the system will be fraught with difficulty. The development process itself is shown, at a high level, in Figure 7.4.

The high-level development process shown in Figure 7.3 indicates that there is a natural break after the first two phases. Once the contextual architecture and the conceptual architecture are agreed upon and signed off on, then work on the later phases can begin, with considerable parallel working. However, it is difficult to make useful progress on the later stages until these first two are fairly fully defined. The temptation to go straight to an implementation of certain products and tools should be avoided, since this is the source of so many severe problems during the operational phase.

It is also important not to be confused by the positioning of the subprocess "Define Operational Security Architecture." The operational security architecture itself cuts across all of the other five layers, but the development process for that operational security architecture is best delayed until after the contextual and conceptual security architectures have been defined and signed off on.

7.1.8 SABSA Life Cycle

The SABSA life cycle (Figure 7.5) is designed to align with the IT life cycle. Regardless of the scope, the SABSA framework provides a structured approach for successful delivery. It can address the challenges of service management, enterprise-wide architecture, designing the infrastructure for a new business initiative, implementing a single IT project, or complying with governance and compliance directives, and can provide a road map to achieve the objectives.

In the SABSA life cycle, the first two phases of the SABSA development process are grouped into an activity called "Strategy and Concept." This is followed by an activity called "Design," which embraces the design of the logical, physical,

	Assets (What)	Motivation (Why)	Process (How)	People (Who)	Location (Where)	Time (When)
Contextual	Business Requirements Collection; Information Classification	Business Risk Assessment; Corporate Policy Making	Business-driven Information Security Management Program	Business Security Organization Management	Business Field Operations Program	Business Calendar & Timetable Management
Conceptual	Business Continuity Management	Security Audit, Corporate Compliance; Metrics, Measures & Benchmarks; SLAs	Change Control; Incident Management; Disaster Recovery	Security Training; Awareness; Cultural Development	Security Domain Management	Security Operations Schedule Management
Logical	Information Security; System Integrity	Detailed Security Policy Making; Policy Compliance Monitoring; Intelligence Gathering	Intrusion Detection; Event Monitoring; Security Process Development Security Service Management; System Development Controls; Configuration Management	Access Control; Privilege & Profile Administration	Applications Security Administration & Management	Applications Deadline & Cutoff Management
Physical	Database Security; Software Integrity	Vulnerability Assessment; Penetration Testing; Threat Assessment	Rule Definition; Key Management; ACL Maintenance; Backup Administration; Computer Forensics; Event Log Administration; Antivirus Administration	User Support; Security Help Desk	Network Security Management; Site Security Management	User A/C Aging; Password Aging; Crypto Key Aging; Administration of AccessControl Time Windows
Component	Product & Tool security & Integrity	Threat Research; Vulnerability Research; CERT Notifications	Product Procurement; Project Management; Operations Management	Personnel Vetting; Supplier Vetting; User Administration	Platform Workstation & Equipment Security Management	Time-out Configuration; Detailed Security Operations Sequencing

Figure 7.3. SABSA framework for security management.

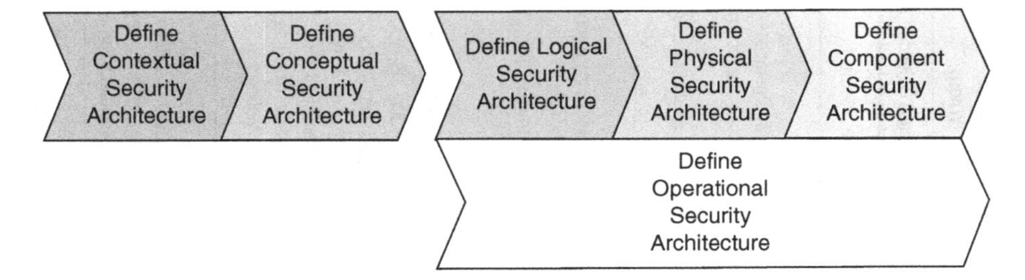

Figure 7.4. The SABSA development process.

component and operational architectures. The third activity is "Implement," followed by "Manage and Measure." The significance of the "Measure" activity is that early in the process, target performance metrics are developed, as shown in the attributes section below. Once the system is operational, it is essential to measure actual performance against targets, and to manage any deviations observed. Such management may simply involve the manipulation of operational parameters, but it may also feed back into a new cycle of development.

7.1.9 SABSA Attributes

A further refinement is the use of SABSA business attributes. These attributes are compiled from extensive experience with numerous clients in many countries and industry sectors. Over the course of that work, it became apparent that although every business is unique, there are commonly recurring themes. This experience has been used to create a taxonomy of SABSA business attributes, shown in Figure 7.6. These are organized under seven group headings.

It should also be noted that each of the identified attributes (and possibly others) represents, in the positive aspect, a desirable business "virtue," and in the negative aspect, a risk. As such, the attributes that are relevant to a particular organization also provides a basis for developing control objectives to manage risks.

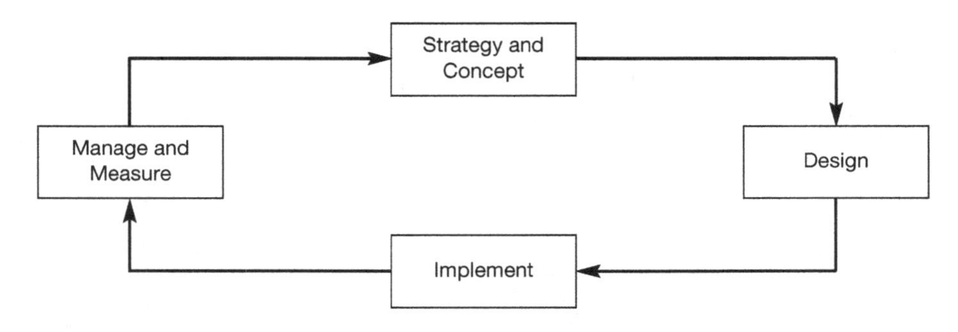

Figure 7.5. The SABSA life cycle.

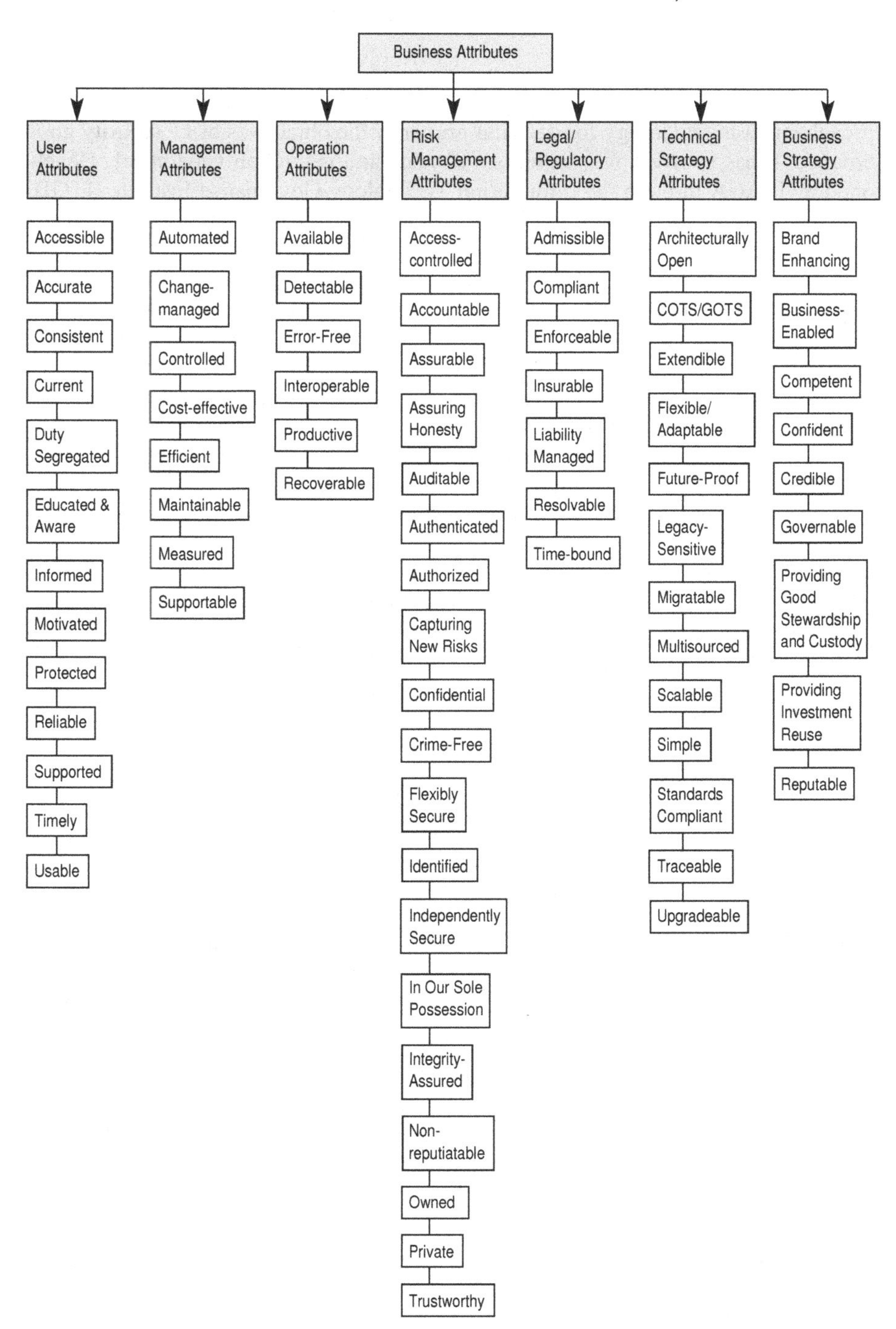

Figure 7.6. SABSA business security attributes.

7.2 CobiT

CobiT is a well-developed, comprehensive system that can provide both an approach and a methodology for defining primarily the objectives of IT security governance. It has evolved over a number of years and is now on version 4.1. Developed by ISACA through the Information Technology Governance Institute (ITGI), which until 2003 was an association of only IT auditors, CobiT has at its core an audit perspective, although it is becoming increasingly inclusive as it evolves.

Since 2003, ISACA has developed the Certified Information Security Manager Program with a perspective focused on information security management. Although the two approaches are in many ways complementary, there are fundamental differences.

It should be noted that CobiT can be used by itself or can be utilized within the broader framework of the SABSA architecture to address a number of the identified attributes involving control objectives. Although its focus is primarily on IT, it is possible to cover most information security requirements as well since most information security issues directly or tangentially involve information technology.

The basic sequence for determining control objectives relevant to both CobiT and ISO 27001 is to utilize risk assessments and analysis to determine what risks need to be mitigated and to what extent. This is then translated into control objectives to address those risks. CobiT offers many tools and well-defined approaches and techniques for IT security management and, to a considerable extent, the information security manager as well.

To quote the ITGI:

> COBIT is based on the analysis and harmonisation of existing IT standards and good practices and conforms to generally accepted governance principles. It is positioned at a high level, driven by business requirements, covers the full range of IT activities, and concentrates on what should be achieved rather than how to achieve effective governance, management and control. Therefore, it acts as an integrator of IT governance practices and appeals to executive management; business and IT management; governance, assurance and security professionals; and IT audit and control professionals. It is designed to be complementary to, and used together with, other standards and good practices.
>
> Implementation of good practices should be consistent with the enterprise's governance and control framework, appropriate for the organisation, and integrated with other methods and practices that are being used. Standards and good practices are not a panacea. Their effectiveness depends on how they have been implemented and kept up to date. They are most useful when applied as a set of principles and as a starting point for tailoring specific procedures. To avoid practices becoming shelfware, management and staff should understand what to do, how to do it and why it is important.
>
> To achieve alignment of good practice to business requirements, it is recommended that COBIT be used at the highest level, providing an overall control framework based on an IT process model that should generically suit every enterprise. Specific practices and standards covering discrete areas can be mapped up to the COBIT framework, thus providing a hierarchy of guidance materials. [3]

The focus for CobiT is defining IT control objectives and developing the controls to meet them. The COBIT framework sets out 34 processes to manage and control information and the technology that supports it (Table 7.2). The processes are divided into four domains:

1. **Plan and Organize**—This domain covers strategy and tactics, and concerns the identification of the way IT can best contribute to the achievement of the business objectives. Furthermore, the realization of the strategic vision needs to be planned, communicated, and managed for different perspectives. Finally, a proper organization as well as technological infrastructure must be put in place.
2. **Acquire and Implement**—To realize the IT strategy, IT solutions need to be identified, developed, or acquired, as well as implemented and integrated into the business process. In addition, changes to and maintenance of existing systems are covered by this domain to make sure that the life cycle is continued for these systems.
3. **Deliver and Support**—This domain is concerned with the actual delivery of required services, which range from traditional operations over security and continuity aspects to training. To deliver services, the necessary support processes must be set up. This domain includes the actual processing of data by application systems, often classified under application controls.
4. **Monitor and Evaluate**—All IT processes need to be regularly assessed over time for their quality and compliance with control requirements. Thus, this domain addresses management's monitoring and evaluation of IT performance and increased control, ensuring regulatory compliance and providing IT governance oversight.

7.3 CAPABILITY MATURITY MODEL

The Capability Maturity Model (CMM) broadly refers to a process improvement approach based on a process model. CMM also refers specifically to the first such model, developed by the Software Engineering Institute (SEI) in the mid-1980s, as well as the family of process models that followed. A process model is a structured collection of practices that describe the characteristics of those practices proven by experience to be effective. CMM can be used to assess an organization against a scale of five process maturity levels. Each level ranks the organization according to its standardization of processes in the subject area being assessed. The subject areas can be as diverse as software engineering, systems engineering, project management, risk management, system acquisition, information technology services, and personnel management.

CMM is based on a description of the attributes of five levels of "maturity" of organizational structures and processes. A number of variations have been developed based on the original. While quite descriptive, easy to use, and almost intuitive, the approach is fairly subjective.

Table 7.2. CobiT high-level control objectives

	Plan and Organize
PO1	Define a strategic IT Plan and direction
PO2	Define the information architecture
PO3	Determine technological direction
PO4	Define the IT processes, organization, and relationships
PO5	Manage the IT Investment
PO6	Communicate management aims and direction
PO7	Manage IT human resources
PO8	Manage quality
PO9	Assess and manage IT risks
PO10	Manage projects
	Acquire and Implement
AI1	Identify Automated solutions
AI2	Acquire and maintain application software
AI3	Acquire and maintain technology infrastructure
AI4	Enable operation and use
AI5	Procure IT resources
AI6	Manage changes
AI7	Install and accredit solutions and changes
	Deliver and Support
DS1	Define and manage service levels
DS2	Manage third-party services
DS3	Manage performance and capacity
DS4	Ensure continuous service
DS5	Ensure systems security
DS6	Identify and allocate costs
DS7	Educate and train users
DS8	Manage service desk and incidents
DS9	Manage the configuration
DS10	Manage problems
DS11	Manage data
DS12	Manage the physical environment
DS13	Manage operations
	Monitor and Evaluate
ME1	Monitor and evaluate IT processes
ME2	Monitor and evaluate internal control
ME3	Ensure regulatory compliance
ME4	Provide IT governance

Source: Wikipedia.

ISACA has developed a version of CMM used in CobiT based on the following six levels:

0. Nonexistent—Organization does not recognize the need for information security

1. Ad-hoc—Risks are considered on an ad-hoc basis and no formal processes exist.
2. Repeatable but intuitive—There is an emerging understanding of risk and the need for security
3. Defined process—Company-wide risk management policies and security awareness
4. Managed and measurable—Risk assessments are standard procedure. Roles and responsibilities are assigned. Policies and standards have been developed.
5. Optimized—Organization-wide processes implemented, monitored, and managed

Although software development was the original focus of CMM, the approach works for any set of processes. National Institute of Standards and Technology (NIST) uses a modified version of CMM as well, which is available on their website (http://www.nist.gov/).

CMM can be used in several ways. One is as a measure of the relative maturity levels of various processes or security activities to see if they meet desired objectives. The levels can be used to define KGIs as well. They can also be used for setting information security objectives. For example, CMM Level 4, Managed and Measurable, can be used as a straightforward set of objectives with high-level characteristics and attributes for developing a security strategy.

A detailed description of each of the six ISACA maturity levels follows.

0—Nonexistent

- Risk assessment for processes and business decisions does not occur. The organization does not consider the business impacts associated with security vulnerabilities or development project uncertainties. Risk management has not been identified as relevant to acquiring IT solutions and delivering IT services.
- The organization does not recognize the need for information security. Responsibilities and accountabilities are not assigned for ensuring security. Measures supporting the management of information security are not implemented. There is no information security reporting and no response process to information security breaches. There is a complete lack of a recognizable system security administration process.
- There is no understanding of the risks, vulnerabilities, and threats to IT operations, or the impact of loss of IT services to the business. Service continuity is not considered as needing management attention.

1—Initial/Ad Hoc

- The organization considers IT risks in an ad hoc manner, without following defined processes or policies. Informal assessments of project risk take place as determined by each project.
- The organization recognizes the need for information security, but security awareness depends on the individual. Information security is addressed on a

reactive basis and not measured. Information security breaches invoke "finger pointing" responses if detected, because responsibilities are unclear. Responses to information security breaches are unpredictable.

- Responsibilities for continuous service are informal, with limited authority. Management is becoming aware of the risks related to and the need for continuous service.

2—Repeatable but Intuitive

- There is an emerging understanding that IT risks are important and need to be considered. Some approach to risk assessment exists, but the process is still immature and developing.
- Responsibilities and accountabilities for information security are assigned to an information security coordinator with no management authority. Security awareness is fragmented and limited. Information security information is generated but not analyzed. Security tends to respond reactively to information security incidents by adopting third-party offerings, without addressing the specific needs of the organization. Security policies are being developed, but inadequate skills and tools are still being used. Information security reporting is incomplete, misleading, or not pertinent.
- Responsibility for continuous service is assigned. The approaches to continuous service are fragmented. Reporting on system availability is incomplete and does not take business impact into account.

3—Defined Process

- An organization-wide risk management policy defines when and how to conduct risk assessments. Risk assessment follows a defined process that is documented and available to all staff through training.
- Security awareness exists and is promoted by management. Security awareness briefings have been standardized and formalized. Information security procedures are defined and fit into a structure for security policies and procedures. Responsibilities for information security are assigned but not consistently enforced. An information security plan exists, driving risk analysis and security solutions. Information security reporting is IT-focused, rather than business-focused. Ad hoc intrusion testing is performed.
- Management communicates consistently the need for continuous service. High-availability components and system redundancy are being applied piecemeal. An inventory of critical systems and components is rigorously maintained.

4—Managed and Measurable

- The assessment of risk is a standard procedure and exceptions to following the procedure would be noticed by IT management. It is likely that IT risk management is a defined management function with senior-level responsibility. Senior management and IT management have determined the levels of risk that the organization will tolerate and have standard measures for risk/return ratios.

- Responsibilities for information security are clearly assigned, managed, and enforced. Information security risk and impact analysis is consistently performed. Security policies and practices are completed with specific security baselines. Security awareness briefings have become mandatory. User identification, authentication, and authorization are standardized. Security certification of staff is established. Intrusion testing is a standard and formalized process leading to improvements. Cost/benefit analysis, supporting the implementation of security measures, is increasingly being utilized. Information security processes are coordinated with the overall organization security function. Information security reporting is linked to business objectives.
- Responsibilities and standards for continuous service are enforced. System redundancy practices, including use of high-availability components, are consistently deployed.

5—Optimized

- Risk management has developed to the stage where a structured, organization-wide process is enforced, followed regularly and managed well.
- Information security is a joint responsibility of business and IT management and is integrated with enterprise security business objectives. Information security requirements are clearly defined, optimized, and included in a verified security plan. Security functions are integrated with applications at the design stage and end users are increasingly accountable for managing security. Information security reporting provides early warning of changing and emerging risk, using automated active monitoring approaches for critical systems. Incidents are promptly addressed with formalized incident response procedures supported by automated tools. Periodic security assessments evaluate the effectiveness of implementation of the security plan. Information on new threats and vulnerabilities is systematically collected and analyzed, and adequate mitigating controls are promptly communicated and implemented. Intrusion testing, root cause analysis of security incidents, and proactive identification of risk form the basis for continuous improvements. Security processes and technologies are integrated organizationwide.
- Continuous-service plans and business-continuity plans are integrated, aligned, and routinely maintained. Buy-in for continuous service needs is secured from vendors and major suppliers.

7.4 ISO/IEC 27001/27002

This standard and code of practice can serve to provide an approach to security governance, although, to some extent by inference. That is, 27001 is a *management* system with a focus on control objectives, not a strategic governance approach. The linkage between control objectives and strategic business objectives is not explicitly addressed. 27001 is a standard that can be certified to, whereas 27002 is a code of security practice that cannot. CobiT has been mapped to 27001 and 27002, and they cover much of the same functional territory, although organized differently. Additionally, CobiT covers IT governance extensively.

Many organizations and security managers find benefits in utilizing international standards and it might be useful to utilize a higher-level framework or other approach for setting strategic security governance objectives that then feed into the ISO model.

7.4.1 ISO 27001

The ISO 27001 standard published in October 2005 replaced the original BS7799-2 standard. It is the specification for an Information Security Management System (ISMS). BS7799 itself was a long-standing standard, first published in the 1990s as a code of practice. As it matured, the second part emerged to cover management systems against which certification can be granted.

ISO 27001 added to and enhanced the content of BS7799-2 and harmonized it with other standards. The objective of the standard is to "provide a model for establishing, implementing, operating, monitoring, reviewing, maintaining, and improving an Information Security Management System" [4].

Adoption should be a strategic decision since "The design and implementation of an organization's ISMS is influenced by their needs and objectives, security requirements, the process employed, and the size and structure of the organization."

The standard defines its "process approach" as "The application of a system of processes within an organization, together with the identification and interactions of these processes, and their management." It employs the PDCA (Plan–Do–Check–Act) model (Figure 7.7) to structure the processes, and reflects the principles set out in the OECG Guidelines of the OECD [5].

ISO 27001 contains Annex A, which is a table of 134 controls under various headings derived from ISO 17799 (now 27002), listing basic minimum controls that are generally applicable to most organizations. It also requires that any of these controls not implemented must have stated reasons for the exclusion.

Although it is oriented toward information security management, ISO 27001 does make reference to business linkages in general terms. The following is the introduction from ISO 27001, containing guidelines for establishing the Information Security Management System and the scope and coverage of the standard. Section numbers refer to other parts of the standard.

Establish the ISMS

The organization shall do the following:

a) Define the scope and boundaries of the ISMS in terms of the characteristics of the business, the organization, its location, assets and technology, including details of and justification for any exclusions from the scope (see 1.2).

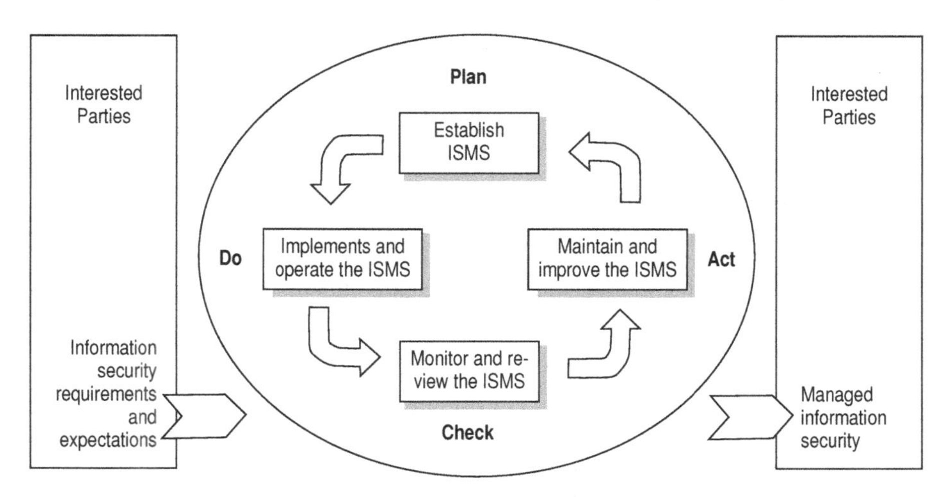

Plan (establish the ISMS)	Establish ISMS policy, objectives, processes, and procedures relevant to managing risk and improving information security to deliver results in accordance with an organization's overall policies and objectives.
Do (implement and operate the ISMS)	Implement and operate the ISMS policy, controls, processes, and procedures.
Check (monitor and review the ISMS)	Assess and, where applicable, measure process performance against ISMS policy, objectives, and practical experience, and report the results to management for review.
Act (maintain and improve the ISMS)	Take corrective and preventive actions, based on the results of the internal ISMS audit and management review or other relevant information, to achieve continual improvement of the ISMS.

Figure 7.7. PDCA model applied to ISMS.

b) Define an ISMS policy in terms of the characteristics of the business, the organization, its location, assets, and technology that:
   1. Includes a framework for setting objectives and establishes an overall sense of direction and principles for action with regard to information security;
   2. Takes into account business and legal or regulatory requirements, and contractual security obligations;
   3. Aligns with the organization's strategic risk management context in which the establishment and maintenance of the ISMS will take place;
   4. Establishes criteria against which risk will be evaluated (see 4.2.1c); and
   5. Has been approved by management.

c) Define the risk assessment approach of the organization.
   1. Identify a risk assessment methodology that is suited to the ISMS and the identified business information security, legal and regulatory requirements.

   2. Develop criteria for accepting risks and identify the acceptable levels of risk. (see 5.1f).
      The risk assessment methodology selected shall ensure that risk assessments produce comparable and reproducible results.

d) Identify the risks.
   1. Identify the assets within the scope of the ISMS, and the owners of these assets.
   2. Identify the threats to those assets.
   3. Identify the vulnerabilities that might be exploited by the threats.
   4. Identify the impacts that losses of confidentiality, integrity, and availability may have on the assets.

e) Analyze and evaluate the risks.
   1. Assess the business impacts upon the organization that might result from security failures, taking into account the consequences of a loss of confidentiality, integrity, or availability of the assets.
   2. Assess the realistic likelihood of security failures occurring in the light of prevailing threats and vulnerabilities, and impacts associated with these assets, and the controls currently implemented.
   3. Estimate the levels of risks.
   4. Determine whether the risks are acceptable or require treatment using the criteria for accepting risks established in 4.2.1c.2.

f) Identify and evaluate options for the treatment of risks.
   Possible actions include:
   1. Applying appropriate controls;
   2. Knowingly and objectively accepting risks, providing they clearly satisfy the organization's policies and the criteria for accepting risks (see 4.2.1c.2);
   3. Avoiding risks; and
   4. Transferring the associated business risks to other parties, for example, insurers, suppliers.

g) Select control objectives and controls for the treatment of risks. Control objectives and controls shall be selected and implemented to meet the requirements identified by the risk assessment and risk treatment process. This selection shall take account of the criteria for accepting risks (see 4.2.1c.2) as well as legal, regulatory, and contractual requirements. The control objectives and controls from Annex A shall be selected as part of this process as suitable to cover the identified requirements. The control objectives and controls listed in Annex A are not exhaustive and additional control objectives and controls may also be selected.
   NOTE: Annex A contains a comprehensive list of control objectives and controls that have been found to be commonly relevant in organizations. Users of this International Standard are directed to Annex A as a starting point for control selection to ensure that no important control options are overlooked.

h) Obtain management approval of the proposed residual risks.

i) Obtain management authorization to implement and operate the ISMS.

j) Prepare a Statement of Applicability. A Statement of Applicability shall be prepared that includes the following:

1. The control objectives and controls selected in 4.2.1g and the reasons for their selection;
2. The control objectives and controls currently implemented (see 4.2.1e.2); and
3. The exclusion of any control objectives and controls in Annex A and the justification for their exclusion. [6]

7.4.2 ISO 27002

ISO 27002 is the renamed ISO 17799 Code of Practice for Information Security. It basically provides a large number of potential controls and control mechanisms, which may be implemented, in theory, subject to the guidance provided within ISO 27001.

The Code of Practice established "guidelines and general principles for initiating, implementing, maintaining, and improving information security management within an organization." The actual controls listed in the standard are intended to address the specific requirements identified via a formal risk assessment. The standard is also intended to provide a guide for the development of "organizational security standards and effective security management practices, and to help build confidence in inter-organizational activities."

The areas covered by 27002 include:

1. Risk assessment
2. Security policy—management direction
3. Organization of information security—governance of information security
4. Asset management—inventory and classification of information assets
5. Human resources security—security aspects for employees joining, moving within, and leaving an organization
6. Physical and environmental security—protection of computer facilities
7. Communications and operations management—management of technical security controls in systems and networks
8. Access control—restriction of access rights to networks, systems, applications, functions, and data
9. Information systems acquisition, development, and maintenance—building security into applications
10. Information security incident management—anticipating and responding appropriately to information security breaches
11. Business continuity management—protecting, maintaining, and recovering business-critical processes and systems
12. Compliance—ensuring conformance with information security policies, standards, laws, and regulations

Within each section, information security controls and their objectives are specified and outlined. The information security controls are generally regarded as good practices for achieving those objectives. For each of the controls, implementation guidance is provided. Specific controls are not mandated since:

1. Each organization is expected to undertake a structured information security risk assessment process to determine its specific requirements before selecting controls that are appropriate to its particular circumstances. The introduction section outlines a risk assessment process, although there are more specific standards covering this area such as ISO Technical Report TR 13335 GMITS, Part 3—Guidelines for the Management of IT Security—Security Techniques; and BS 7799 Part 3.
2. It is practically impossible to list all conceivable controls in a general-purpose standard. Industry-specific implementation guidance for ISO/IEC 27001 and 27002 are anticipated to give advice tailored to organizations in the telecommunications, financial services, healthcare, lotteries, and other industries.

7.5 OTHER APPROACHES

The National Cyber Security Summit Task Force Corporate Governance Report of 2002 [7] provides a comprehensive set of actions required for security governance and can serve as a detailed basis for determining the desired state as well as objectives. It is summarized below for consideration as a model and framework.

7.5.1 National Cyber Security Task Force, Information Security Governance: A Call to Action

The road to information security goes through corporate governance. America cannot solve its cyber security challenges by delegating them to government officials or CIOs. The best way to strengthen U.S. information security is to treat it as a corporate governance issue that requires the attention of Boards and CEOs.

The Corporate Governance Task Force was formed in December 2003 to develop and promote a coherent governance framework to drive implementation of effective information security programs. Although information security is often viewed as a technical issue, it is also a governance challenge that involves risk management, reporting, and accountability. As such, it requires the active engagement of executive management.

Today's economic environment demands that enterprises in both the public and private sectors reach beyond traditional boundaries. Citizens, customers, educators, suppliers, investors, and other partners are all demanding more access to strategic resources. As enterprises reinvent themselves to meet this demand, traditional boundaries are disappearing and the premium on information security is rising. Heightened concerns about critical infrastructure protection and homeland security are accelerating this trend.

In this report, we provide a framework and guidelines to help organizations assess their performance and put in place an information security governance program. By themselves, however, these tools are not enough. To succeed, we need a private sector commitment to implement this framework and begin to integrate information security into their corporate governance programs.

As we embrace information security governance, it is important to remember that, like quality, it is a journey that requires continuous improvement over time. We are still in the early stages of this journey. As we progress, we will not only reap the rewards of productivity growth, customer satisfaction, and improved competitiveness, but also gain the larger reward of enhanced homeland security. We encourage you to join us in this effort.

1. Introduction and Purpose

1.1. This document provides an overview of the various elements of an information security governance program. It is based on the October 2003 Business Software Alliance report, "Information Security Governance—Toward a Framework for Action." Information security governance is a subset of good organizational governance, which comprises the set of policies and internal controls by which organizations are directed and managed.

1.2. The purpose of this document is to provide a comprehensive framework for ensuring the effectiveness of information security controls over information resources, to provide effective management and oversight of the related information security risks, to provide for development and maintenance of minimum controls required to protect an organization's information and information systems, and to provide a mechanism for oversight of the information security program.

1.3. Recognizing that it is not practical to account for every organization type, size, and structure in a single framework, this document is offered in a general form that can be adapted to most organizational structures. The organization leaders must adapt the framework elements to their specific situation, assigning functions to those staff members most capable and appropriate. Additional guidance is offered to assist with this task in companion documents.

1.4. As used in this document, the term information security means protecting information and information systems from unauthorized use, disclosure, disruption, modification, or destruction to provide the following:

- Confidentiality, which means preserving an appropriate level of information secrecy
- Integrity, which means guarding against improper information modification or destruction, and includes ensuring information nonrepudiation and authenticity
- Availability, which means ensuring timely and reliable access to and use of information

2. Responsibilities of the Board of Directors/Trustees
The board of directors/trustees or similar governance entity should provide strategic oversight regarding information security, including:

2.1. Understanding the criticality of information and information security to the organization.

2.2. Reviewing investment in information security for alignment with the organization strategy and risk profile.

2.3. Endorsing the development and implementation of a comprehensive information security program.

2.4. Requiring regular reports from management on the program's adequacy and effectiveness.

3. Responsibilities of the Senior Executive
The Senior Executive, typically a Chief Executive Officer accountable to the Board of Directors or like entity, should provide oversight of a comprehensive information security program for the entire organization, including:

3.1. Assigning the responsibility, accountability and authority for each of the various functions described in this document to appropriate individuals within the organization.

3.2. Overseeing organizational compliance with the requirements of this document, including through any authorized action to enforce accountability for compliance with such requirements.

3.3. Reporting to the board of directors/trustees or similar governance entity on organization compliance with the requirements of this document, including:
- A summary of the findings of evaluations, with an indication of the level of residual risk deemed acceptable
- Significant deficiencies in organization information security practices
- Planned remedial action to address such deficiencies.

3.4. Designating an individual to fulfill the role of senior information security officer, who should possess professional qualifications, including training and experience, required to administer the information security program as defined in this document, and head an office with the mission and resources to assist in pursuing organizational compliance with this document.

4. Responsibilities of the Executive Team Members
Specific members of the executive team, typically those managers reporting directly to the Senior Executive, should oversee the organization's security policies and practices, including:

4.1. Overseeing the development and implementation of policies, principles, standards, and guidelines on information security, consistent with the guidance of accepted security practices such as ISO/IEC 17799, and in Section 7 of this document.

4.2. Seeing that information security management processes are integrated with organization strategic and operational planning processes.

4.3. Coordinating information security policies and procedures with related information resources management policies and procedures.

4.4. Providing information security protections commensurate with the risk and magnitude of the harm resulting from unauthorized use, disclosure, disruption, modification, or destruction of information collected or maintained, or information systems used or operated by or on behalf of the organization.

4.5. Seeing that each independent organizational unit develops and maintains an information security program.

4.6. Seeing that the senior information security officer, in coordination with organizational unit heads, reports periodically to the Senior Executive on the effectiveness of their information security program, including the progress of remedial actions.

4.7. Seeing that the Senior Information Security Officer assists organizational unit managers concerning their information security responsibilities.

5. Responsibilities of Senior Managers
The head of each independent organizational unit should see that senior organizational unit managers provide information security for the information and information systems that support the operations and assets under their control, including through:

5.1. Assessing the risk and magnitude of the harm that could result from the unauthorized use, disclosure, disruption, modification, or destruction of such information or information systems.

5.2. Implementing policies and procedures that are based on risk assessments and cost-effectively reduce information security risks to an acceptable level.

5.3. Determining the levels of information security appropriate to protect such information and information systems.

5.4. Periodically testing and evaluating information security controls and techniques to see that they are effectively implemented.

5.5. Seeing that the organization has trained personnel sufficient to assist the organization in complying with the requirements of this document and related policies, procedures, standards, and guidelines.

5.6. Seeing that all employees, contractors, and other users of information systems are aware of their responsibility to comply with the information security policies, practices, and relevant guidance appropriate to their role in the organization.

6. Responsibilities of All Employees and Users
All employees of an organization and, where relevant, third-party users share responsibilities for the security of information and information systems accessible to them, including:

6.1. Awareness of the information security policies, practices and relevant guidance appropriate to their role in the organization.

6.2. Compliance with the security policies and procedures related to the information and information systems they use.

6.3. Reporting of vulnerabilities or incidents affecting security or security policy compliance to the appropriate management channels.

7. Organizational Unit Security Program
Each independent organizational unit should develop, document, and implement an information security program, consistent with the guidance of accepted security practices such as ISO/IEC 17799, to provide information security for the information and information systems that support the operations and assets of the organizational unit, including those provided or managed by another organizational unit, contractor, or other source, which includes:

7.1. Periodic assessment of the risk and magnitude of the harm that could result from the unauthorized use, disclosure, disruption, modification, or destruction of such information or information systems.

7.2. Policies and procedures that are based on risk assessments and cost-effectively reduce information security risks to an acceptable level.

7.3. Seeing that information security is addressed throughout the life cycle of each information system.

7.4. Pursuing compliance with the requirements of this document, policies, and procedures as may be prescribed by the Senior Executive, and any other applicable legal, regulatory, or contractual requirements.

7.5. Subordinate plans for providing adequate information security for networks, facilities, and systems or groups of information systems, as appropriate.

7.6. Security awareness training to inform personnel, including contractors and other users, of information systems who support the operations and assets of the organizational unit, of:

- Information security risks associated with their activities
- Their responsibilities in complying with organization policies and procedures designed to reduce these risks

7.7. Periodic testing and evaluation of the effectiveness of information security policies, procedures, and practices, to be performed with a frequency depending on risk, but no less than annually.

7.8. A process for pursuing remedial action to address any deficiencies in the information security policies, procedures, and practices.

7.9. Procedures for detecting, reporting, and responding to security incidents, including:

- Mitigating risks associated with such incidents before substantial damage is done
- Notifying and consulting with a federal or industry information security incident center
- Notifying and consulting with the corporate disclosure committee, law enforcement agencies, or other companies or organizations in accordance with law

7.10. Plans and procedures to pursue continuity of operations for information systems that support the operations and assets of the organization.

8. Organizational Unit Reporting
Each independent organizational unit should:

8.1. Report periodically to the appropriate senior executive on the adequacy and effectiveness of the information security program, including compliance with the requirements of this document.

8.2. Address the adequacy and effectiveness of the information security program in the organizational unit's budget, investment, and performance plans and reports.

8.3. Report any significant deficiency in organizational information security practices, planned remedial actions to address such deficiencies, and an indication of the level of residual risk deemed acceptable.

8.4. In consultation with the appropriate senior executive, report as part of the performance plan a description of the time periods, and the resources, including budget, staffing, and training, that are necessary to implement the information security program elements required.

8.5. Provide customers and business partners with timely notice and opportunities for comment on proposed information security policies and procedures to the extent that such policies and procedures affect communication with them.

9. Independent Information Security Program Evaluation
Although not practical for all organization types and sizes, each independent organizational unit should perform a regular evaluation to validate the effectiveness of its information security program.

9.1. Each evaluation by the organizational unit under this section could be performed by an internal auditor or an independent external auditor, and it should include:

- Testing of the effectiveness of information security policies, procedures, and practices of a representative subset of the organizational unit's information systems
- An assessment of compliance with the requirements of this document and related information security policies, procedures, standards, and guidelines

9.2. The evaluation recommended by this section

- Should be performed in accordance with generally accepted auditing standards
- May be based in whole or in part on an audit, evaluation, or report relating to programs or practices of the applicable organizational unit

9.3. Organizational units and evaluators should take appropriate steps to ensure the protection of related information, which, if disclosed, may adversely affect information security. Such protections should be commensurate with the risk and comply with all applicable laws and regulations.

9.4. The Senior Executive should summarize the results of the evaluations conducted under this section in a report to the Board of Directors/Trustees, or a similar governance entity in which such an entity exists. [7]

REFERENCES

1. Pouchard, Mark, Meta Group, ZDNET, 11/11,02, http://techupdate.ZDNET.com, 2002.
2. Sherwood Applied Business Security Architecture, SABSA Institute, 2005.
3. CobiT 4.1 Executive Summary Framework, ITGI, 2007.
4. International Organization for Standardization, ISO 27001, 2005.
5. Organization for Economic Co-operation and Development, Open Ethics Compliance Group Committee Guidelines, 2000.
6. ISO/IEC 27001:2005(E), 2005.
7. National Cyber Security Summit Task Force, Corporate Governance Report, 2002.

Chapter **8**

Risk Management Objectives

A critical component of identifying the "desired state" of security is to determine rather precisely the overall risk management objectives. It is treated separately from general security objectives since it forms the basis and rationale for all security activities. These objectives must be defined with sufficient clarity that the manager of risk has a point of reference to know what to manage to and to design the controls to achieve those objectives.

One of the outcomes for information security governance discussed earlier refers to managing risk "appropriately." This outcome can only be useful for setting objectives to the extent that we determine what the organization considers "appropriate." A persuasive argument can be made that it should mean the optimal point at which the cost of remediation is equal to the cost of compromise, as shown in Figure 8.1. The reality in most organizations is generally far from this point however.

This is exemplified by the TJX case where the loss of over 46 million credit card records to hackers hinged upon the management decision not to spend relatively minor resources to change over from WEP wireless encryption to the more robust WPA, even with the knowledge and warning of the fact that they were highly vulnerable to compromise.

One of the issues that must be addressed in determining risk management objectives is the fact that everyone in the organization has some level of responsibility for managing risk. This means that consideration must be given to whether it is possible to change the enterprise culture if that is what is required to manage risk "appropriately." In other words, objectives must be realistic if they are to be realized. Since it is usually the case that implementing general controls to mitigate risk to acceptable levels will encounter a level of resistance, that can represent a formidable challenge. Though not impossible, it will require senior management buy-in and unwavering support, which in turn will typically require, at a minimum, a very persuasive business case.

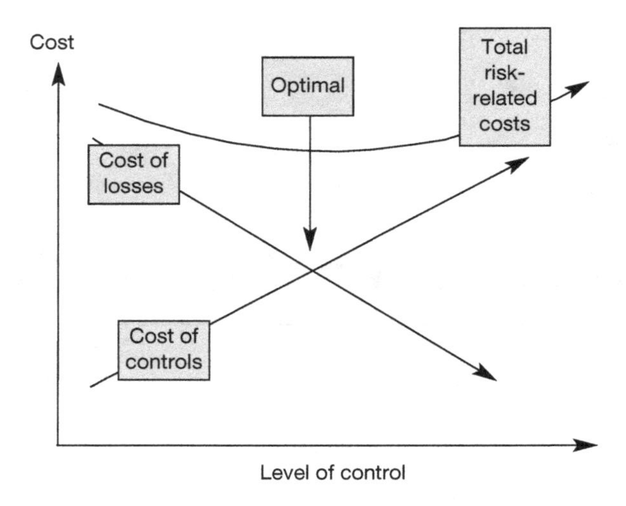

Figure 8.1.

8.1 RISK MANAGEMENT RESPONSIBILITIES

The responsibilities for risk management are well described in the ISACA 2008 Certified Information Security Manager Review Manual:

> The objective of this job practice area (risk management) is to ensure that the information security manager understands the importance of risk management as a tool for meeting business needs and developing a security management program to support these needs. While information security governance defines the links between business goals and objectives and the security program, security risk management defines the extent of protection that is prudent based on business requirements, objectives, and priorities.
>
> The objective of risk management is to identify, quantify, and manage information security-related risks to achieve business objectives through a number of tasks utilizing the information security manager's knowledge of key risk management techniques. Since information security is one component of enterprise risk management, the techniques, methods and metrics used to define information security risks may need to be viewed within the larger context of organizational risk. As indicated in chapter 1 [of the ISACA Manual], information security risk management also needs to incorporate human resource, operational, physical and environmental risks. [1]

8.2 MANAGING RISK APPROPRIATELY

There are several approaches to establishing to a reasonable extent the level of risk management is willing to accept, although this is likely to require several iterations and consideration of cost. Although a great deal of management focus is balancing risks and rewards in a variety of other aspects, many managers still remain relatively unfamiliar with information security risks and consequences. These issues are likely to be more attentively received after the completion of a full and formal risk assessment backed by business impact assessment (BIA) and analysis. This serves to educate management as to the nature of the issues and the magnitude of possible impacts. It also reduces the possibility of someone subsequently pleading ignorance of the risks and protects the security manager.

It is reasonable to assume that had the risks and impacts of the choices resulting in over 250 million dollars in direct losses at TJX been convincingly exposed, it would have resulted in prudent senior management addressing the issue of wireless encryption and protection of credit card information.

One method to arrive at concrete objectives for risk management is to reach agreement on what constitutes acceptable impact in both financial terms on an annualized basis and potentially damaging compromises from a reputation standpoint. Depending on the type of business and the nature of senior management, the relevance of reputational damage will vary considerably. Breaches at universities are the most common but, typically, the educational culture is more inclined to view restrictive controls as more unacceptable than compromises, and it appears not to be a significant hindrance to their operations.

Financial institutions, on the other hand, are often more concerned about erosion of customer trust that direct financial loss and are generally at the other end of the spectrum. The point is that impacts must be considered from the perspective of the particular organization. It also appears that breaches and compromises have become so common that the level of reputational damage and embarrassment of any particular event has become far less onerous; the only thing that would be worse for a manager than an embarrassing security failure would be to be the only one suffering it.

In any event, gaining management consensus on the levels and types of impacts that are acceptable and the ones that must be addressed provides a starting point for the security manager to determine the objectives of risk management. It provides the basis for starting to define the outcome of the "appropriate" level of risk management.

8.3 DETERMINING RISK MANAGEMENT OBJECTIVES

The acceptable levels of impact will need to be specified in various dimensions if they are to be used for determining the "appropriate" management of risk. The same process described in Chapter 7 for describing the desired state of information security will also be necessary to describe the desired state of risk, which, in turn, will define the risk management requirements and objectives.

For example, if we are using the SABSA attribute table shown in Figure 7.6 , there is the operational attribute of "available." Availability is largely a function of system uptime. Defining the "desired" uptime allows analysis of the elements that adversely affect it and what risks must be managed to achieve it. Another attribute in the SABSA table under risk management is "auditable." The risks that must be managed will be things like inadequate logging, incomplete records, poor access control, and so on.

If we are using CMM Level 4, Managed and Measurable, for setting security objectives, the fifteen stated attributes and characteristics can be used in much the same way to determine risk management objectives. The fifteen are:

1. The assessment of risk is a standard procedure, and exceptions to following the procedure would be noticed by security management.
2. Information security risk management is a defined management function with senior-level responsibility.
3. Senior management and information security management have determined the levels of risk that the organization will tolerate and have standard measures for risk/return ratios.
4. Responsibilities for information security are clearly assigned, managed and enforced.
5. Information security risk and impact analysis is consistently performed.
6. Security policies and practices are completed with specific security baselines.
7. Security awareness briefings have become mandatory.
8. User identification, authentication, and authorization are standardized.
9. Security certification of staff is established.
10. Intrusion testing is a standard and formalized process leading to improvements.
11. Cost–benefit analyses, supporting the implementation of security measures, are increasingly being utilized.
12. Information security processes are coordinated with the overall organization security function.
13. Information security reporting is linked to business objectives.
14. Responsibilities and standards for continuous service are enforced.
15. System redundancy practices, including use of high-availability components, are consistently deployed.

These CMM practices are at a fairly high level and will need to be further refined. Objectives defining the desired state of risk management needs to include methods and processes to provide assurance that each target is achieved at the defined level and maintained.

8.3.1 Recovery Time Objectives

Another approach to determining the level of risk acceptable to management is to develop recovery time objectives (RTOs). These will require a business impact assessment (BIA) and analysis to establish the cost of the loss of resources over time. This is a normal requirement for business continuity and disaster recovery planning (BCP/DRP) as well as effective incident management, and is performed in organizations with more mature information security programs as a matter of course.

Since providing shorter recovery times usually results in more expensive solutions (e.g., hot site versus cold site), the level of expense that management will incur to provide assurance of achievable RTOs is an indicator of the level of risk they are willing to accept and consider "appropriate." Of course, the willingness to fund the recovery from an event will be influenced by the likelihood of the occurrence. If the probability of a serious event requiring recovery is low, the level of financial commitment is likely to be as well. The question that may need to be answered is whether a supportable basis exists for the likelihood of requiring full recovery operations or whether the more supportable position will be for better incident management and response.

Since a major security compromise can threaten the existence of the organization, the question for management is to what extent are they willing to "bet the business" that such an eventuality will not occur. From a justification perspective, severe security compromise resulting in a catastrophe will probably be considered at the same level as the possibility of a major earthquake, being hit by a tornado, or fire destroying the premises. The level of mitigation measures for these types of events, including the amount of insurance carried by the organization and its cost, can serve as a good indicator of what constitutes "acceptable" risk and the level of "appropriate" risk management.

Whether management perspective on acceptable risk is ultimately correct will be determined in the fullness of time. The best course of action for the security manager may be to make the implicit explicit and state the obvious, that is, promote the notion that risks from security events receive the same level of support, management buy-in, and explicit acceptance provided to other types of risks with similar potential consequences.

REFERENCE

1. Information Security And Control Association, *2008 CISM Review Manual.*

Chapter 9

Current State

We have defined the "desired state" of security and risk management to achieve the governance outcomes discussed in Chapter 1, including:

- Strategic alignment
- Risk management
- Business process assurance/convergence
- Value delivery
- Resource management
- Performance measurement

The next consideration will be to determine the current state of each of these elements using the identical methods that were used in defining the desired state. By identifying the current state of the identical characteristics and attributes, the gap between current state and the desired state can be evaluated. Although the concept is straightforward, determining the gap is not the same as closing it, which may pose some formidable challenges.

There are always reasons why the current state is current and there are always staunch supporters of the status quo. What needs to be done may appear obvious and yet it may be virtually unachievable given organizational perceptions, culture, financial elements, and other constraints. The vision of what should be is nevertheless useful and the current state can always be improved to some extent. Heading these improvements in the right direction is the purpose of governance. These are aspects that will be considered in Chapter 11.

9.1 CURRENT STATE OF SECURITY

To gauge the current state of the attributes and characteristics of information security governance, use one or more of the approaches described in Chapter 7, including:

- SABSA security architecture
- CobiT
- Capability Maturity Model
- ISO/IEC 27001, 27002
- National Cyber Security Summit Task Force Corporate Governance Framework

9.1.1 SABSA

Using SABSA, the choices of relevant attributes from the chart (Figure 7.6, also listed in Appendix A) that have been selected as a part of describing the desired state must be assessed as to the extent to which they describe the current state. For example if we look at Risk Management Attributes, the second attribute is "Accountable." It is likely that the desired state will include full accountability as a part of defined roles and responsibilities. The current state may be deficiencies in accountability, which could be expressed on an arbitrary scale of 1 to 10 for each functional role, department, or other unit.

Under Business Strategy Attributes is the attribute of "Competent." We will need to identify the competency we are assessing and defining in the desired state and then utilize some measure that can be consistently applied. The SABSA attributes and possible metrics are detailed in Appendix A. The balance of the attributes can be selected from the table in the Appendix, which includes ways to measure them.

9.1.2 CobiT

The 34 high-level control objectives can also be used to determine current state of information security, although the focus is on IT security. In most instances, IT can be replaced by IS and still be relevant. For example, in the Plan and Organize domain, PO1 states, "Define a Strategic IT Plan and Directions." The current state is likely to be binary, that is, a strategic security plan either exists or it does not.

PO8 is "Manage Quality" and we may need to define what areas of quality we are concerned with. From an information security governance and management perspective, this may be the quality of security services (QOSS) or it may be using ISO 9000 or a Six Sigma approach to process quality with a target of virtually error-free security operations and processes. However, It is important to remember that a high quality of deficient processes are not likely to achieve the desired results.

9.1.3 CMM

With CMM, the descriptions of each of the six levels will provide an approximation of the current level of the maturity of security in an organization. The approach lacks granularity but can serve as the rough measure for each of the six outcomes defined. For example, if the security objective, or desired state, is to achieve a Level 4 CMM, Managed and Measurable, and we want to consider the outcome of Strategic Alignment, statements 11, 12, and 13 address this issue:

11. Cost–benefit analyses, supporting the implementation of security measures, are increasingly being utilized.
12. Information security processes are coordinated with the overall organization security function.
13. Information security reporting is linked to business objectives.

Cost–benefit analysis considers benefits in terms of supporting business objectives and the current state may be that this is not performed and only ranks 1 or 2 on the CMM scale. There may be no or little coordination of security processes and security reporting does not consider business objectives and likewise is low on the scale.

Appropriate risk management is addressed by CMM Level 4, statements 1, 2, 3, 4, and 5.

Business process assurance/convergence is covered by 4 and12.

Value delivery by statement 11 and 12.

Resource management by 4, 12, and 15.

Performance measurement by statements 3, 5, 6, and 13.

Because CMM granularity is low, the extent to which the statements apply may need to be stated in terms of a percentage of achievement. For example, statement 11 is "cost benefit analysis, supporting the implementation of security measures, are increasingly being utilized." The current state may be that it occurs in 10% of situations and the objective might be 100%. Analysis can then determine under what conditions it occurs and when it does not.

9.1.4 ISO/IEC 27001, 27002

The ISO standards are more suited to developing a management approach than determining strategic outcomes using the ISMS processes and control objectives in ISO/IEC 27001; and the code of practice and controls in ISO/IEC 27002. Nevertheless, it can provide useful governance guidance and can be effectively used to establish the current state of security. Just as with CobiT, the management criteria and the 134 common required controls identified can be used to determine if they are currently employed or not.

9.1.5 Cyber Security Task Force Governance Framework

The governance framework defined by the Task Force arguably identifies virtually all the required elements of providing adequate information security governance. By using it as a checklist, the current state of organizational security can be described in considerable detail.

For example, item 3 of the framework states:

> **3. Responsibilities of the Senior Executive.** The Senior Executive, typically a Chief Executive Officer accountable to the Board of Directors or like entity, should provide oversight of a comprehensive information security program for the entire organization, including:
>
> 3.1. Assigning the responsibility, accountability and authority for each of the various functions described in this document to appropriate individuals within the organization.
>
> 3.2. Overseeing organizational compliance with the requirements of this document, including through any authorized action to enforce accountability for compliance with such requirements.
>
> 3.3. Reporting to the board of directors/trustees or similar governance entity on organization compliance with the requirements of this document, including
>
> - A summary of the findings of evaluations, with an indication of the level of residual risk deemed acceptable
> - Significant deficiencies in organization information security practices
> - Planned remedial action to address such deficiencies.
>
> 3.4. Designating an individual to fulfill the role of senior information security officer, who should possess professional qualifications, including training and experience, required to administer the information security program as defined in this document, and head an office with the mission and resources to assist in pursuing organizational compliance with this document."

This level of detail can be used to easily identify whether the conditions exist and to what extent. An analysis of the entire listing will provide a good snapshot of the current state of information security governance for most organizations.

9.2 CURRENT STATE OF RISK MANAGEMENT

It will be necessary to develop a comprehensive evaluation of the current state of risk management. Just as with the state of security, it will be necessary to use the methods selected in Chapter 8 to develop a clear descriptive view of current practices and capabilities. To round out the picture of the current state, a full risk assessment will be required as well.

9.3 GAP ANALYSIS—UNMITIGATED RISK

The next step will be a complete analysis for each element, attribute, and characteristic to determine the gap between the current state and the desired state. At this juncture, as each element is evaluated in terms of the difference between the current and desired states, consideration must be given to what it will require to "fill" the gap if one exists. That is, will it simply take more or will it require different processes, technologies, controls, and so on.

Assuming the desired state defines the state of security that manages risk to acceptable levels, then, the "gap" between the current state and desired state is the unmitigated risk that implementing the strategy is designed to address.

For an example, let us consider some samples of each approach and how a gap analysis might be conducted.

9.3.1 SABSA

Let us examine the following SABSA security attributes taken from the User Attributes in the table in Appendix A that might have been used (along with others) to determine both current and desired state.

For the desired state attribute, "Protected—the user's information and access privileges should be protected against abuse by other users or by intruders," the current state attribute might be, "User's information is protected by user ID and passwords with no requirement for change and no automated approach to ensure good password construction." The gap is fairly obvious and remediation is as well, although it may not necessarily be easy to convince the enterprise that the more restrictive controls are necessary.

For the desired state attribute, "Reliable—the services provided to the user should be delivered at a reliable level of quality," the current state attribute might be, "System performance is inconsistent and customers report seeing other people's accounts on occasion. Customer data is frequently corrupted and unreadable." The gap here is also apparent and is a technical problem that governance has not addressed. It may be that operational standards are inadequate or not enforced.

For the desired state, "Responsive—the users obtain a response within a satisfactory period of time that meets their expectations," the current state might be, "Users often complain that security is slow to respond to their issues and cause delays in necessary system changes and project implementation." The gap is obvious, although the causes are not. It may be a lack of security staff and resources, or it may be that review processes are cumbersome and inefficient.

9.3.2 CMM

As an example for CMM, let us examine some of the 15 statements used for defining current and desired states.

The first statement for defining the desired state for CMM Level 4, Managed and Measurable, is "1. The assessment of risk is a standard procedure, and exceptions to following the procedure would be noticed by security management." An example of a current state familiar to many security practitioners might be, "Assessment of risk is ad-hoc and there is no formal process requiring notification of management. Scope and frequency of assessments are not mandated and the activity is discouraged by IT management as time consuming and costly."

In this example, several deficiencies are clear and filling the gap will require several things be addressed. The first is the attitude of IT management. Depending on the organization, its structure, culture, and position of the security manager, addressing this issue will require various activities and may or may not be achievable. For the objective to be realized will require either gaining IT cooperation by education or pursuasion, or else senior management support mandating the requirement as a matter of policy.

Policies and standards may need to be created or modified and approved as necessary for the regular performance of risk assessments and their scope as well as analysis and reporting requirements.

The next CMM element to consider is, "2. Information security risk management is a defined management function with senior-level responsibility." The current state in many organizations is likely to be, "Information security is a part of IT reporting to the CIO. The position is a low-level manager with responsibilities for technical security of desktop computers and the data center."

The position as a manager in the IT department is structurally deficient and prone to conflicts of interest. Information security is a regulatory function with a focus on safety; IT is focused on performance. The regulator cannot report to the regulated and the pressure for increased performance and reduced costs for IT is often at the expense of security in these situations. In heavily regulated sectors such as financial institutions, there are specific prohibitions against this reporting structure in many jurisdictions recognizing the inherent problem with the practice. To close this gap will require senior management to recognize the issue and address it by elevating security management to a senior position reporting to the COO, CEO, or, perhaps, to the audit committee of the board. The scope, authority and resources of security management must be equivalent to any other major operational unit to meet this objective and achieve effective security governance.

When the desired state is, "3. Senior management and information security management have determined the levels of risk that the organization will tolerate and have standard measures for risk/return ratios," the current state might be, "Levels of acceptable risk have not been defined and risk management is a reactive process with mitigation considered after an event is determined to be excessively damaging to the organization. Cost/benefit analysis is not performed for security activities and no metrics are utilized to determine performance levels."

To address these deficiencies, several elments must be addressed. A process must be developed to ensure that management determines some parameters for acceptable risk in order for security management to know what level to manage risk to—how much is enough, too little, or too much. The fact that risk is not in effect being managed but is determined by the extent to which the organization suffers damage can be considered a lack of due care and is utterly fundamental to the responsibilities of senior management and boards. The failure to perform analysis and implement performance metrics means that there is no rational basis to prioritize allocation of limited resources and virtually no information on which to base security management decisions.

As shown in the foregoing examples, gap analysis, when completed, can paint a reasonably complete picture of existing deficiencies and provide the basis for determining the activities required to close the gaps.

Chapter **10**

Developing a Security Strategy

Strategic business or organizational goals are the long term objectives essential for setting the direction of an organization. They can be considered the overarching purpose for the organization and are often stated in abstract aspirational terms such as "better living through chemistry". Whereas in simpler days strategic goals were more about creating wealth for the stakeholders, now that seems too simplistic and loftier aspirations are the norm if not necessarily the practice. Regardless, somewhere within every organization exist notions about the things it must achieve to be successful.

Assuming that organizational goals and strategic security governance objectives have been defined as discussed in Chapter 5, Section 5.1 and the current and desired states have been developed, the next issue will be what is needed to develop and implement an information security strategy to achieve the objectives.

Strategy is another common term with many definitions that do little to clarify its meaning. For our purposes, the original military meaning is the most useful: Strategy is the plan to achieve an objective. An information security strategy is, therefore, the objectives of information security coupled with the plans to achieve it. The concept of design is nearly synonymous with strategy. Both require an objective and a plan, although a strategy implies actions as well, whereas design does not.

One common feature of strategy that should be considered before embarking on the arduous efforts of developing one is that they often fail, sometime spectacularly. It is evident that history is littered with examples of bad strategies and that situation continues unabated. The ongoing rate of (partial or complete) failure for IT projects is in the 70% range, which can be largely attributed to inadequate design or strategy, poor execution, and insufficient metrics. This is despite the fact that business strategy has been extensively studied for more than four decades by a branch of psychology called behavioral economics, which studies, among other things, decision making processes and departure from rational choice. Departure from rational choice is evident in all aspects of society and highly visible in the periodic irrational exuberance of the stock and real estate markets.

A recent report from McKinsey and Co. [1] merits consideration as well. It cautions that often the "approach to strategy involves the mistaken assumption that a predictable path to the future can be paved from the experience of the past." It goes on to suggest that strategic outcomes can not be predetermined given today's turbulent business environment. As a result, McKinsey proposes to define strategy as "a coherent and evolving portfolio of initiatives to drive shareholder value and long-term performance." This change in thinking requires management to develop a you are what you do perspective as opposed to you are what you say. In other words, companies are defined by the initiatives they prioritize and drive, not merely by mission and vision statements.

10.1 FAILURES OF STRATEGY

An review of behavioral economics would require a separate volume, however, some of the findings of the common causes of the failures of strategy can be illuminating and useful for an organization embarking on the development of a security strategy. These are typical and common pitfalls at all levels of the typical organization based on normal psychological traits of humans. The purpose for considering them here is to gain the awareness to compensate appropriately for natural tendencies and avoid the pitfalls of departure from rational choice.

The causes of strategy failures include some obvious reasons such as inadequate or faulty analysis, greed, unmitigated ambition, and other corporate malfeasance, but other causes are not so well understood. Some of the more common ones include the following:

Overconfidence. This includes excessive confidence in the ability to make accurate estimates. A common manifestation is that most people are reluctant to estimate a wide ranges of possible outcomes and prefer to be "precisely wrong rather than vaguely right." Findings include the observation that most tend to be overconfident of their own abilities. For organizational strategies based on accurate assessments of core capabilities, this can be particularly troublesome.

Optimism. People tend to be optimistic in their forecasts. A combination of overconfidence and overoptimism can have a disastrous impact on strategies based on estimates of what may happen. Typically, these estimates are unrealistically precise and overly optimistic.

Anchoring. Research shows that once a number has been presented to someone, a subsequent estimate of even a totally unrelated subject involving numbers will "anchor" on the first number. Though potentially useful for marketing purposes, anchoring can have serious consequences in developing strategies when future outcomes are anchored in past experiences.

Status quo bias. There is a strong tendency to stick with familiar approaches even when they are demonstrably inadequate or ineffective. "The devil you know is better than the one you don't" illustrates this penchant. Research also indicates that concern with loss is stronger than excitement about possible gain.

Endowment effect. This is a similar bias for people to keep what they own and that simply owning something makes it more valuable to the owner.

Mental accounting. Defined as the inclination to categorize and treat money differently depending on where it comes from, where it is kept, and how it is spent. Mental accounting is common even in the boardrooms of conservative and otherwise rational corporations. Some examples of this include being less concerned with expenses against a restructuring charge than those against the profit and loss; imposing cost caps on a core business while spending freely on a start-up; and creating new categories of spending, such as "revenue-investment" or "strategic investment," as if the name impacted the amounts or justification.

Herding instinct. It is a fundamental human trait to conform and seek validation by the actions of others. This can be observed by the "faddism" in security as evidenced by the sudden adoption and deployment across industries of identity management or intrusion detection. It is based in the fear of being left out and missing the boat. Executives often make decisions based on what everybody else is doing. It is aptly demonstrated by one pundit who quipped, "For senior managers the only thing worse than making a huge strategic mistake is being the only person in the industry to make it." The implications for strategy development are clear.

False consensus. The tendency for people to overestimate the extent that others share their views, beliefs, and experiences.

Confirmation bias. Seeking only those opinions and facts that support one's own beliefs.

Selective recall. Remembering the facts and experiences that reinforce current assumptions.

Biased evaluation. Acceptance of evidence supporting the preferred hypotheses while challenging and rejecting contradictory evidence; often accompanied by charging critics with hostile motives and impugning their competence,

Group Think. The pressure for agreement in a team-based or consensus-oriented culture. False consensus can lead to ignoring or minimizing important threats or weaknesses in plans and persisting with doomed strategies.

10.2 ATTRIBUTES OF A GOOD SECURITY STRATEGY

At this juncture, we know the current state and the desired state; the strategy will be the plan to get from here to there. It will be the road map and define the interplay of resources and constraints, of people, processes, technologies, and requirements.

To develop a workable strategy, it will also be necessary to know the culture and the landscape, that is, the context. The context will determine what is possible and the kinds of solutions and approaches that will be acceptable to the organization and the environment it operates in. The resources that are available must be known and the constraints must be understood as well.

What is the hallmark of a good strategy? Are there elements we can consider likely to result in a successful strategy? In Section 10.1 we discussed some of the ways that strategies fail as a result of common psychological tendencies. A review of the list may be useful to counteract normal tendencies that may distort plans but there will be other aspects to consider as well.

1. Are the objectives realistic and achievable?
This may be another approach to considering the excessive optimism and overconfidence mentioned in Section 10.1. It is not uncommon for security practitioners with ambitious goals for implementing effective security to be frustrated by the difficulties in gaining management support and adequate resources. An information security strategy cannot rest solely on the assumption that every objective will be substantially achieved. It is more prudent to consider how to optimize partial successes and how the various components of the strategy can still function to some degree. Any part of a plan that requires the complete success of a prior aspect should be revisited and acceptable alternatives devised.

2. Are the objectives likely to achieve the desired outcomes?
Developing an effective security strategy will require priorities and trade-offs. For example, consideration must be given to which security attributes take precedence when conflicting requirements exist. A typical example would be the trade-offs between performance and safety. Not that they cannot coexist to a considerable extent but if a particular business requirement for high throughput or fast response conflicts with the processing necessary for high assurance of identity, which will take priority? An example would be when organization using PKI authentication for high-value transactions is unable to check the Certificate Revocation List (CRL) on a Certificate Authority (CA) because the server was unavailable. Would completing the transaction take priority over the risk that the certificate of the other party was not valid? In general, the issues of confidentiality, integrity, and availability must be considered both from a strategy and controls perspective in terms of trade-offs and focus.

Strategy is similar to design and, just as in the SDLC approach, requires the same notions of developing requirements and specifications:

- Establishing requirements
- Determining feasibility
- Solution architecture and design
- Proof of concept
- Full development and construction
- Integration testing

- Deployment
- Quality and acceptance testing
- Maintenance
- Systems' end of life

According to the ISACA:

> A good security strategy should address and mitigate risks while complying with the legal, contractual and statutory requirements of the business as well as provide demonstrable support for the business objectives of the organization and maximize value to the stakeholders. The security strategy also needs to address how the organization will embed good security practices into every business process and area of the business. Often, those charged with developing a security strategy think in terms of controls as the means to implement security. Controls, while important, are not the only element available to the strategist. In some cases, for example, reengineering a process can reduce or eliminate a risk without the need for controls. Or, potential impacts may be mitigated by architectural modifications rather than controls. It should also be considered that, in some cases, mitigating risks can reduce opportunities to the extent of being counterproductive.
>
> Ultimately, the goal of security is business process assurance, regardless of the business. While the business of a government agency may not result directly in profits, it is still in the business of providing cost-effective services to its constituency and must still protect the assets for which it has custodial care. Whatever the business, its primary operational goal is to maximize the success of business processes and minimize impediments to those processes.
>
> Some might argue that the primary goal of security is to protect information assets. However, information is only an asset insofar as it supports the primary purpose of the business, generating revenues (or cost-effective services) through value-add processes. All other information is a liability. As some organizations have discovered, information that should have been subject to a retention and destruction policy turned out to be a major liability when incriminating e-mails were discovered by the opposition in a lawsuit. Even if not incriminating, useless information consumes resources and is a liability. [2]

10.3 STRATEGY RESOURCES

The development of a strategy will require a detailed understanding of the resources available to implement it. There will be a number of ways that any particular objective resulting from the gap analysis discussed in Chapter 9, Section 9.3 can be addressed. Note: the "gaps" identified in the analysis constitutes unmanaged risks that the strategy will be designed to address using the available resources within the existing constraints.

Resources are any activity, process, asset, technology, individual, policy, and so on that can be utilized in some manner to move toward addressing a gap (i.e., mitigating a risk) that serves to implement the strategy.

As used here, resources can provide information or be a part of a solution to a problem. Their absence or ineffectiveness may also constitute a risk or threat that must be addressed by the strategy. To "normalize" the understanding of a number of processes and elements of security that suffer from a variety of definitions and differences in understanding globally, each of the identified resources are described in relation to their relevance to a security strategy.

Resources available to implement a strategy will typically include:

- **Policies.** The high-level statement of management intent, expectations, and direction; can be considered to be the "constitution" of security governance.
- **Standards.** Set the allowable boundaries for people, processes, procedures, and technologies necessary to meet the intent of policies; Can be considered the "law" of security governance
- **Procedures.** Procedures are the detailed steps necessary to accomplish a particular task and must conform to the standards
- **Guidelines.** For the purposes of information security, guidelines are helpful narratives in executing procedures including suggestions, tools, and so on.
- **Architecture(s).** Defines the relationships between objects, the information flows between them, and the inputs and outputs, as well as other aspects such as schemas, specifications, metrics, test points, and so on. They can range from contextual and conceptual, to logical, functional, physical, and operational.
- **Controls—physical, technical, procedural.** Controls are any regulatory element, whether process, procedure, technology, or physical component, for example, access controls, procedural controls, and firewalls.
- **Countermeasures.** Countermeasures directly target mitigation of any threat or vulnerability; can be considered a targeted control.
- **Layered defenses.** Layered defenses are the practice of adding subsequent or sequential controls in an effort to ensure that failure of one control will not compromise the entire system.
- **Technologies.** Numerous technology controls are available to address growing threats and losses. These will generally be preventive, detective, corrective, or compensatory in nature, or some combination of these. For example, a firewall can both be preventive and detective, as well as possibly compensatory.
- **Personnel security.** People continue to be the greatest risk to security through overwork, lack of training, carelessness, indifference, accident, mistakes, and, occasionally, malice. An effective strategy will consider the human element as central to implementation and secure operations.
- **Organizational structure.** The structure of the organization can be beneficial to effective security strategy development or a monumental obstacle. Considerations for strategy development must include organizational structures, reporting relationships, real or potential conflicts between various orga-

nizational units, and whether it is a command and control structure or a flat, decentralized structure.

- **Roles and responsibilities.** The extent to which roles and responsibilities are clear and well defined must be considered in strategy development and will have a significant impact on implementation and operation of a security program.
- **Skills.** Strategies that utilize existing skills and proficiencies will be easier to implement and must be assessed in the planning phase.
- **Training.** Major changes to operations as a result of implementing security governance will require either skills acquisition or training or both. Training requirements must be considered in the development of a security governance strategy and the development of controls operation and configuration.
- **Awareness and education.** User awareness and education often provides the greatest return on investment in terms of security. Plans should include provisions for ongoing security awareness using various methods such as computer-based training and security briefings.
- **Audits.** Audits are a normal fact of life for most organizations but are often not utilized to best effect. The strategy should consider how to coordinate with auditors and utilize audits as a force for needed changes in overall governance.
- **Compliance enforcement.** Compliance enforcement is often one of the most problematic elements of security governance and due consideration must be given in the strategy to how policies and standards can be enforced both from a technical perspective and the more difficult physical and procedural standpoint. Effective policy and standards compliance is essential to a successful security program.
- **Threat analysis.** Ongoing evaluation of existing and emerging threats must be planned for in strategy development.
- **Vulnerability analysis.** Technical vulnerability scanning is standard procedure in most organizations but physical and operational vulnerabilities are often not tested and are ignored. The strategy should consider how all these elements will be addressed.
- **Risk assessment.** Risk assessment must be a standard practice at the strategic, management, and operational levels, and must to cover entire business processes from input to output in addition to relevant external factors. Policies and standards must provide the requirements for appropriate assessment of risk on an ongoing basis.
- **Business impact assessment.** Business impact assessments (BIAs) are essential for determining protection and recovery priorities, and should be a policy requirement defined in the strategy.
- **Resource dependency analysis.** Resource dependency analysis can be an alternative to BIAs in determining protection and recovery priorities. It is based on analyzing the resources required for critical business processes.

- **Outsourced security providers.** Outsourced security providers are a viable option for implementing a strategy but carry some attendant risk that must be addressed. The strategy must consider the options for outsourcing as a viable approach to program implementation.
- **Other organizational support and assurance providers.** The strategy must consider how to optimize integration of a variety of other assurance providers in the organization to maximize security cost effectiveness.
- **Facilities.** Facilities management must be considered in a security program strategy given the critical nature of physical and environmental impact on information security effectiveness.
- **Environmental security.** External environmental threats and risks must be addressed by the security program strategy.
- **Metrics and monitoring.** Effective governance is not possible without adequate metrics and suitable monitoring. All key controls must be designed with metrics in mind to ensure their continued operation. Monitoring of key processes and controls is also essential and the strategy must address how this will be accomplished.

10.3.1 Utilizing Architecture for Strategy Development

If major initiatives are anticipated to achieve the defined governance objectives, it is evident that there will be many moving parts that must be considered in terms of dealing with particular aspects as well as the relationship between the elements. It may be beneficial to consider a security architecture to formally address all the required elements as well as their interrelationships and ability to achieve the desired outcomes. If we consider the SABSA approach discussed in Chapter 8, populating each of the 36 elements in the matrix repeated in Figure 10.1 is likely to provide a high level of assurance that this has been achieved.

Each of the resources listed (and perhaps others) will be utilized in one or more of the 36 elements. For example, Policies are in the Motivation column and the Logical layer, technology will populate several of both the Physical and Component layers, and organizational structure will be in the Contextual layer and People column.

10.3.2 Using CobiT for Strategy Development

Mapping available resources to address gaps in the CobiT control objectives is relatively straightforward, as shown in the following examples.

In Planning and Organization (PO),

PO1—Define a Strategic IT Plan and Direction. This can be addressed from resources on the list by architectures, policies, organizational structure, and, of course, by the resource we are developing here, strategy.

	Assets (What)	Motivation (Why)	Process (How)	People (Who)	Location (Where)	Time (When)
Contextual	The Business	Business Risk Model	Business Process Model	Business Organization Relationships	Business Geography	Business Time Dependencies
Conceptual	Business Attributes Profile	Control Objectives	Security Strategies and Architectural Layering	Security Entity Model and Trust Framework	Security Domain Model	Security-Related Lifetimes and Deadlines
Logical	Business Information Model	Security Policies	Security Services	Entity Schema and Privilege Profiles	Security Domain Definitions and Associations	Security Processing Cycle
Physical	Business Data Model	Security Rules, Practices, and Procedures	Security Mechanisms	Users, Applications, and the User Interface	Platform and Network Infrastructure	Control Structure Execution
Component	Detailed Data Structures	Security Standards	Security Products and Tools	Identities, Functions, Actions, and ACLs	Processes, Nodes, Addresses, and Protocols	Security Step Timing and Sequencing
Operational	Assurance of Operational Continuity	Operational Risk Management	Security Service Management and Support	Application and User Management and Support	Security of Sites, Networks, and Platforms	Security Operations Schedule

Figure 10.1. SABSA Matrix

PO4—Define the IT Processes, Organization, and Relationships. Policies, standards, organizational structure, and roles and responsibilities will be resources for PO4.

PO9—Assess and Manage IT Risks. Threat and vulnerability analysis and Risk and impact assessment from our resources list address this requirement.

10.3.3 Using CMM for Strategy Development

Using the CMM 4 Managed and Measurable approach, the available resources must be considered for each of the gaps identified in the 15 statements.

For example, assume that we must address "2—Information security risk management is a defined management function with senior-level responsibility." If this is not the case, the resources that would be relevant to addressing this issue can be Policies and, possibly, Standards, Organizational Structure, and Roles and Responsibilities.

Now consider "3—Senior management and information security management have determined the levels of risk that the organization will tolerate and have standard measures for risk/return ratios."

If this is not the case, resources to address this issue can include Policy and Standards development setting forth the requirement, risk, impact assessment and analysis, development of continuity, disaster plans and the attendant recovery time objectives, and the requirement for business case development, including financial anlysis as a project requirement standard.

Each of the remaining 12 CMM attributes and characteristics must be evaluated in a similar manner to determine what resources are available to address the gaps.

10.4 STRATEGY CONSTRAINTS

There are also a number of constraints that must be addressed when developing a security strategy and subsequent action plan. Constraints will be any condition that diminishes or adversely impacts the achievement of any of the selected attributes. Constraints can be direct, consequential, or peripheral.

A direct constraint is a characteristic that either prohibits a particular solution or renders it ineffective. For example, the strategy may call for greater control of unauthorized intrusions by hackers. One solution, or countermeasure, would be to round up all hackers and shoot them. While it might be effective, it is obviously not feasible and is a prohibited solution. More realistically, prohibitions can be legal, ethical, practical, or a matter of organizational policy. Ineffectiveness can be a problem of excessive control, such as using cannon to swat a fly, or inadequate control, such as solely relying on voluntary policy compliance without monitoring.

Consequential constraints are those for which the application of a control has subsequent effects beyond that intended. An example is assigning long, complex passwords, with the likely consequence that they will be written down and subject to being compromised. Adding substantial work or difficulty to users as a means of control is inviting deleterious effects whose probability must be considered as a constraint to the approach.

Peripheral constraints address the possibility that the intended effects of a particular approach extends beyond the intended targets. Any control, however effective and efficient at addressing a particular security gap that adversely impacts the ability to get required work done, will be subject to peripheral constraints.

There are two general classes of constraints that must be considered: contextual and operational.

10.4.1 Contextual Constraints

These constraints constitute the boundaries or limits within which the strategy must operate. That is, the approaches and solutions to achieving objectives must not fall outside the defined contextual constraints. These constraints can include:

- **Law**—Legal and regulatory requirements
- **Physical**—Capacity, space, and environmental constraints
- **Ethics**—Appropriate, reasonable, and customary
- **Policy**—Must meet the organizational policy mandates
- **Culture**—Both inside and outside the organization
- **Costs**—Time and money
- **Personnel**—Resistance to change, resentment against new constraints
- **Organizational structure**—How decisions are made and by whom; turf protection
- **Resources**—Capital, technology, and people
- **Capabilities**—Knowledge, training, skills, and expertise
- **Time**—Window of opportunity, mandated compliance
- **Risk tolerance**—Threats, vulnerabilities, and impacts

Some of the constraints, such as ethics and culture, may have been dealt with in developing both the desired and current state. Others can arise as a consequence of developing the road map and action plan when unusual solutions are required to address particular issues.

10.4.2 Operational Constraints

These constraints will be the inherent capabilities and limitations of the solutions and approaches themselves in addressing the gaps. For example, policies and standards are important resources for addressing identified "gaps" but absent good compliance will do little to manage the risk. In other words, merely writing something down will not ensure that it is accomplished. Solutions to addressing gaps need to be verifiable and have the capability of providing feedback as to operational status and effectiveness.

An unreliable or inaccurate technical solution may be a greater problem than the one it was intended to remedy. Arguably, bad or incorrect information may be worse than none at all.

Some operational constraints to consider can include:

- **Manageable**—An approach difficult or impossible to manage will be a constraint
- **Maintainable**—The level of maintenance required for controls must be considered
- **Efficient**—A serious lack of control efficiency can be a constraint
- **Effective**—Controls that are not effective will not be a good choice
- **Proportional**—The cost of a control cannot exceed its benefits
- **Reliable**—Any control that is not consistent and dependable will limit benefits
- **Accurate**—The solution must clearly address the defined problem
- **In Scope**—Is the effect of the solution limited to the intended area or can there be unintended consequences?

REFERENCES

1. McKinsey and Co., http://mckinsey.com/client service/strategy/insight.asp.
2. Information Security and Control Association, *2008 CISM Review Manual*.

Chapter **11**

Sample Strategy Development

Combining the elements of the prior chapters, it is possible to construct a complete strategy and the plans to implement it, assuming that various other activities have taken place, such as determining assets and classifying them as to criticality and sensitivity. This chapter goes through the entire sequence with an example of a typical organization using CMM 4 Managed and Measurable as a straightforward approach for setting objectives. Any of the approaches suggested and, undoubtedly, others can be used in exactly the same way. The ten-step process will track a typical organization through gap analysis of current state and desired state to determine the objectives. Then we will evaluate the "gaps," which are the risks that need to be mitigated, and determine control objectives. The control objectives will be those activities that address the risk and bring the current state to the desired state.

The next step is to examine the options for controls to meet those objectives. The controls choices are evaluated against available resources. Then both contextual and operational constraints are evaluated, and controls modified or replaced as needed. Monitoring and metrics, as applicable, must be designed for each control. Control strength is then evaluated to determine effectiveness and reliability to see if additional layering is required to achieve the needed assurance levels. Project plans can then be formulated to implement the controls consistent with available resources, timelines, budgets and budget cycles, and so on. The result will be a road map and action plan to implement the strategy.

For each of the 15 CMM 4 statements, each of the following 10 steps of the strategy development process covered in the prior chapters is provided. The steps include:

1. Define desired state
2. Determine current state
3. Perform gap analysis
4. Set control objectives

Information Security Governance. By Krag Brotby

5. Determine resources
6. Define constraints
7. Evaluate control choices
8. Design controls with available resources
9. Design monitoring and metrics for controls
10. Develop project and management plans

The 15 CMM statements are a high-level culmination of underlying policies and practices, and in using them, it is often necessary to break them down into their constituent parts to determine what resources and constraints may be relevant.

11.1 THE PROCESS

The goal for our example is CMM 4, Managed and Measurable. For most organizations, this provides an adequate level of security but a substantial effort is required to achieve it. The average for financial institutions overall is in the mid- to high-three range and for all organizations, the average falls in the middle-two range. The fifteen stated requirements may suffice for most organizations but can obviously be modified or extended into the optimized CMM Level 5 or customized where needed.

Gap analysis will in many instances be straightforward but it may also be important to understand the circumstances that are the cause of the current state in order to understand constraints and devise achievable mitigation measures.

The following examples of current state will be typical of many organizations based on extensive experience and are used to demonstrate the process for the a sample of the CMM statements.

Desired state:
CMM 1. The assessment of risk is a standard procedure, and exceptions to following the procedure would be noticed by security management.

Current state:
Risk assessments are not performed. Security management reports to IT and IT management contends that they know the risks and risk assessment is an uneccesary use of resources. Management has a limited understanding of security risk and is totally focused on business performance and short-term share value.

Gaps:
- Lack of awareness of necessity for risk assessment
- Inadequate understanding of risk exposures
- Lack of guidance for remediation resource allocation
- Risk assessment is not a requirement of policy

Control objectives:

- Enforced requirement for risk assessment and analysis on a consistent basis
- Security awareness training to achieve understanding of defined security responsibilities throughout the enterprise
- Risk/reward analysis of business activities

Resources:

- Policies—Policy requirement for risk assessments
- Standards—Setting standard for frequency, scope, and approved methodologies
- Audits—Ensuring audit scope includes policies, standards, and compliance
- Regulatory Requirements—Defining and enforcing relevant legal requirments
- Awareness training on security responsibilities—Implementing regular security awareness training
- Job descriptions—Including specific security requirments in job descriptions
- Personnel performance reviews—Including security responsibilities in performance reviews

Constraints:

- Budget—Insufficient resources allocated for regular risk assessment
- Authority—Security management lacks authority to require risk assessments
- Culture—performance-oriented organization with a "we'll handle it when it happens" management style; high risk tolerance based on lack of understanding impacts and probabilities; excessively optimistic
- Personnel—Sales oriented, prudence not rewarded, management of risk discouraged if any performance impact

Evaluate control choices:

Policy and standards development. Defining the requirements for risk assessment and awareness training in policy with supporting standards is necessary if not already existing. Gaining acceptance and/or enforcement may be a problem; however, audit findings and legal or regulatory requirements can provide impetus. Audit reports are usually presented to the board of directors and findings of a lack of due care or potential negligence, or a failure to comply with legal or regulatory requirements, are often effective at driving improvements.

Design controls with available resources:

- Policy—Modify or create policy stating risk assessment requirement
- Standards—Develop standards for risk assessment scope, frequency, and methodology, mitigation requirements, reporting requirements, compliance review, and enforcement
- Awareness testing and monitoring—Utilize existing training resources as available, periodic testing or surveys, security reviews, and performance reviews

Design monitoring and metrics for controls:

- Policy awareness—Periodic quiz
- Risk assessment standard compliance—Audit, security review

Develop project and management plans:

- Develop risk management policy and standards:
  Risk assessment scope, frequency, and methodology
  Risk assessment monitoring and reporting standards
  Response and mitigation requirements for high-risk findings
- Develop risk assessment procedures
- Risk management policies and standards update, creation, and revision project
- Policy and standards maintenance, review, and acceptance procedures

Desired state:
CMM 2. Information security risk management is a defined management function with senior-level responsibility.

Current state:
Information security is a part of IT security, which deals only with operational issues and reports to the IT manager.

Gap analysis:

- Security management is a low-level, primarily technical function that does not have sufficient authority and scope to identify, understand, and deal effectively with an organization's information security risks.

Control objectives:

- Reorganizing the function into a separate department reporting to the CEO, COO, or the board of directors' audit or risk committee, with defined scope and charter to fully address enterprise information risk.

Resources:

- Risk assessment identifying risks beyond IT
- Business impact assessments and external benchmark analysis
- Incident history and causes (internal and external)
- Legal or regulatory requirements (primarily in financial, insurance and defence sector)
- Policies and Standards
- Audit findings

Constraints:

- Perception of information security as a technical compliance function
- Territoriality of company managers

- Budgets and costs
- Concern of business units about increased oversight and new restrictions

Control choices:
- Security governance awareness and training
- Policy and standards
- Roles and responsibilities
- Due care risk management requirements, structures, and procedures
- International standards
- Compliance (ISMS)

Control design:
- Develop risk management policy and standards:
  Risk management and incident response requirements and scope
  Risk management monitoring and reporting standards
- Develop a process for determining and monitoring an organization's risk tolerance
- Risk management policies and standards update, creation, and revision project

Control monitoring and metrics:
- Periodic impact assessment to contain possible losses within acceptable limits
- Audits and security reviews to monitor policy and standards compliance
- Systematic standards evaluation against baseline security requirements

Project and management plans:
- Developing consensus and sign off on acceptable risk or impacts
- Define roles and responsiblities
- Policy and standards development, approval, and publication
- Determine regulatory and legal requirments. Management decision on compliance levels and acceptance of risk.
- Timing and conditions for audits and reviews
- Establisment or modification as needed of information security steering committee

Desired state:
CMM 3. Senior management and information security management have determined the levels of risk that the organization will tolerate and have standard measures for risk/return ratios.

Current state:
Acceptable risk levels have not been established. Perfunctory cost–benefit analyis is performed occasionally for high-ticket items.

Gap analysis:

Absent defined risk tolerance, it is not possible to know to what extent to mitigate and manage risk. Without the reference point of acceptable risk, effective metrics are not possible. Without financial analysis, resource allocation and prioritization is unlikely to be optimal.

Control objectives :

- Defined process for determining risk tolerance
- Risk metrics or indicators relative to reference point of acceptable risk
- Defined cost–benefit analysis standards for risk management activities

Resources:

- Policies and standards
- Risk assessment
- BCP/DR development of RTOs
- Business impact analysis
- Resource dependency analysis
- Insurance
- Audit findings

Constraints:

- Reluctance of executives to commit to poorly understood concepts
- Inherent lack of precision and uncertainty in processes
- Inadequate resources or understanding need for cost–benefit analysis
- Reactive security posture—"deal with it if it happens"

Control choices:

- Policies, standards, and procedures for determining acceptable risk
- Relevant financial models and analysis approaches
- Business impact assessments and process for determining RTOs
- Evaluation process and standards for risk transfer/acceptance/treatment decisions

Control design:

- Develop policy requirement for periodic impact assessment and formal risk acceptance by senior management
- Develop value at risk (VAR) analysis approach for informing management of risk and probability distribution

Control monitoring and metrics:

- Perform business impact assessment for primary lines of business
- Track and publish impact assessment trends and periodic VAR analysis
- Ensure that defined risks are either formally accepted or adequately mitigated

Project and management plans:
- Implement project to capture historical company and sector data for VAR computations
- Develop plans and projects for performing BIAs for baselines and subsequent trends
- Develop processes to compare risks and potential impacts against acceptable levels

Desired state:
CMM 4. Responsibilities for information security are clearly assigned, managed and enforced.

Current state:
IT security responsibilities are loosely defined or not defined in most job descriptions. Security performance is not a part of job reviews. Most employees are not aware of good security practices needed to manage risk.

Gap analysis:
Security responsibilities must be defined for all employees and contractors if compliance is to be expected. Security compliance is not a part of performance reviews to serve as reminders of expectations and ensure accountability. Awareness is not adequate to maintain reasonable security.

Control objectives:
Security responsibilities are defined in all job descriptions and are a standard part of performance reviews. All personnel are aware of responsibilities and practices necessary to manage risks to the enterprise. Managers promote security awareness to staff and provide oversight to ensure compliance.

Resources:
- Policy
- Standards and procedures
- Awareness training
- Security reviews
- Audits
- HR
- Incident monitoring and root cause analysis

Constraints:
- Perception that security is responsibility of only security department
- General awareness of security responsibilities
- Poor "tone at the top," lack of management support

Control choices:
- Policy, standards, and procedures
- Security awareness training
- Process to include security compliance in performance reviews

Control design:
- Develop supporting policies, standards, and procedures
- Develop cost-effective means of communicating message and ensuring understanding
- Require HR to implement security requirements in performance reviews

Control monitoring and metrics:
- Periodic awareness quiz
- Security reviews
- Audits
- Incident root cause analysis

Project and management plans:
- Policy and standards development, review, and acceptance
- Assign responsibilities for implementation and execution of awareness program

Desired state:
CMM 5. Information security risk and impact analysis is consistently performed.

Current state:
Risk and impacts are not assessed or analyzed except on an ad hoc basis or as a result of an incident

Gap analysis:
Risk and impact assessment are fundamental elements required to steer risk management efforts and the basis and rationale for all security activities. Security efforts remain typically reactive and resource allocation is rarely related to risk/reward ratios.

Control objectives:
Risk and impact assessments are consistent procedures based on standards. Scope is defined and standards for reporting are determined. Severity criteria are developed and mitigation requirements are defined for each severity level.

Resources:
- Policy, standards, and procedures
- Risk and impact assessments
- Other organizational support and assurance providers

Constraints:
- Culture
- Costs
- Resources
- Capabilities

Control choices:
- Policy, standards, and procedures
- Risk and impact assessments

Control design:
- Development of appropriate policy, standards, and procedures review and acceptance process

Control monitoring and metrics:
- Use trending as a metric
- Monitoring of scope and findings of assessments

Project and management plans:
- Manage policy, standards, and procedures development using KGIs and KPIs
- Develop reliable feedback loop for all assessment processes

Each of the remaining CMM statements can be handled in a similar way to gain an understanding of what the strategy or strategies must address and the tools and constraints necessary for achieving the objectives. Often, the issues that pose the greatest difficulties will be cultural and structural. Though they are challenging, addressing them is not impossible, although it may take considerable effort and time. Clarifying these issues is an essential first step in finding possible approaches and illuminating them for management consideration.

Chapter 12

Implementing Strategy

A strategy, whether presented as a multilayered architecture or a series of defined control objectives and controls designed to achieve the desired state, must be translated into a series of actionable items. This can be a sequence of projects or initatives or a combination of activities in a security program.

The strategy serves to guide the actions and activities of the overall security program; in other words, the basis for the rules and structures necessary for governance. The articulation of the strategy comprises the policies that define management intent, expectations, and direction.

12.1 ACTION PLAN INTERMEDIATE GOALS

Consideration for how the organization operates must always be considered. Budget cycles, profitability, and many other issues will affect the implementation of a security program, especially if it is a major initiative.

Existing organizational processes and services must always be used when possible. This will minimize cost and help avoid turf battles.

A formal strategy may cover five years and will need to be carved into reasonably sized chunks commensurate with the normal organizational cycles. This will require a set of intermediate goals consistent with budgets and other cyclical factors, and, of course, consistent with the overall objectives.

Parts of the program will have dependencies that must be considered. Standards cannot be developed until policies are completed. Procedures cannot be finalized until standards are completed. Tactical events may also derail the best laid plans and require redirected efforts. When possible, priority should be given to those activities providing the greatest short-term benefits. Early "wins" will help sustain enthusiasm for the overall program as will demonstrable benefits.

12.2 ACTION PLAN METRICS

Consideration must be given to program development metrics. These will typically be project management metrics showing such things as progress against plan and cost against budget.

12.3 REENGINEERING

Controls tend to aggregate over time and there is typically no process to terminate them even when they are no longer effective. Typically, these controls become simply the "way we do things here" and the underlying reasons for their existence may be long forgotten. A prime example is the case of the U.S. Army installation that for many decades had a guard posted in front of the flagpole in front of headquarters. A curious reporter doing a story on the base enquired about the unusual practice and was informed that had always been the case. The reporter tacked down the retired general who had commanded the post many years before who explained with a chuckle that the flagpole had been freshly repainted and a senior officer had brushed against it getting paint on his uniform. Irritated, he ordered a guard to be posted to avoid a similar incident. More than fifty years later, that control remained in place.

Similarly, many controls that were once relevant can become useless over time. Business process reengineering can be used as an approach to assess and analyze control procedures and practices to determine if they can be modified or eliminated. It is not unusual to find that half of all controls can be eliminated and others improved substantially in terms of user convenience, effectiveness, and efficiency.

12.4 INADEQUATE PERFORMANCE

Bad performance of people, processes, and technology in the majority of situations has been shown to be the result of bad structural design. The application of solid architectural approaches and sound engineering principles can usually do a great deal to address this problem.

12.5 ELEMENTS OF STRATEGY

The first elements to consider in implementing strategy will be policies based on the objectives and, subsequently, standards to develop the policies into operational requirements and limits. The development sequence is:

- Determine desired outcomes consistent with business objectives
- Define objectives to achieve the outcomes

- Develop strategy to achieve the objectives
- Develop or modify policies to formalize management intent, expectations, and direction based on the objectives
- Develop standards based on the policies to formalize the rules and boundaries of acceptable activities for people, processes, and technology
- Ensure the development of procedures to provide detailed instructions for accomplishing tasks consistent with standards and capture available process knowledge

12.5.1 Policy Development

Policies must be owned by senior management and form the foundation for governance. They are the primary tools for implementing a security strategy. Strategy drives the policies, which become operational controls through the development of standards. As shown in Chapter 11, in developing a sample strategy each objective of security must be supported by policy. Policies must provide the critical linkage to overall corporate governance and business objectives. They provide the basis for the management of risk necessary to achieve effective information security governance.

Despite the fact that there has been a great deal written about policies and their development, there continues to be considerable confusion about what they are and how they should be constructed, resulting in a lack of consistency in approach, content, and presentation. It is common to see no distinction made between policies, standards, procedures, and guidelines, and they are often intermingled in an awkward amalgam. The result is often voluminous "shelf ware" compiled over time, usually in response to some event or concern in an attempt to provide a specific rule for every eventuality.

A more rational approach is to consider what policies should accomplish and where in the organization ownership of them must lie. It bears repeating that policies are the highest level directives for an organization and, as an expression of management intent, expectations, and direction, must be straightforward, clear, and overarching. Policies must be owned by the governing body and senior management and have a specific business purpose to justify their existence.

12.5.1.1 Attributes of Good Policies

As high-level management statements, policies should rarely be longer than a sentence or two. Even in large organizations, there is rarely any reason for more than two dozen security policies. Each policy should state one mandate in plain language.

There is often a tendency to attempt to create specific or detailed directions using a policy statement, but this is better accomplished by standards or procedures. A sample of an access control policy might read:

> Access to XYZ premises and resources shall be controlled in a manner that effectively precludes unauthorized access.

This statement provides no detail or methodology; rather, it clearly states the intent and expectation and covers all access, whether physical or logical, tangible or intangible.

12.5.1.2 Sample Policy Development

In Chapter 11 we used CMM Level 4, Managed and Measurable, for developing security objectives and controls. The same fifteen stated attributes and characteristics can also provide the basis for policy development in support of the strategy. To demonstrate policy development, a sample policy for each of the fifteen attributes might be as follows.

CMM 1. The assessment of risk is a standard procedure, and exceptions to following the procedure would be noticed by security management.

Policy:
Risk to XYZ Corporation shall be assessed using a standardized approach on a regular basis, or as changes if circumstances warrant.

CMM 2. Information security risk management is a defined management function with senior-level responsibility.

Policy:
Roles and responsibilities for managing risk shall be defined for XYZ Corporation under the direction of an executive-level individual reporting to the Chief Executive.

CMM 3. Senior management and information security management have determined the levels of risk that the organization will tolerate and have standard measures for risk/return ratios.

Policy:
Information security risks shall be managed to defined levels consistent with classification levels and controlled by appropriate security baselines set forth in the related XYZ Corporation Information Security Standards. Acceptable levels of risk shall be defined in terms of maximum acceptable impacts and reviewed and approved by senior management no less than annually or more often as changing circumstances dictate.

CMM 4. Responsibilities for information security are clearly assigned, managed and enforced.

Policy:
Roles and responsibilities of XYZ Corporation shall be unambiguously defined and all required security functions formally assigned to ensure accountability. Acceptable performance shall be ensured by appropriate monitoring and metrics.

CMM 5. Information security risk and impact analysis is consistently performed.

Policy:
Risk and impact analysis shall be required for all critical or sensitive corporate activities and key controls on a periodic basis and as a required part of new intiatives and change management.

CMM 6. Security policies and practices are completed with specific security baselines.

Policy:
Comprehensive policies and standards shall be developed, implemented, maintained, and enforced utilizing appropriate processes to review, monitor, and measure compliance.

CMM 7. Security awareness briefings have become mandatory.

Policy:
All personnel shall be made aware of relevant policies and standards annually or as changes warrant. Proficiency and competence shall be assessed on a regular basis and appropriate training provided as needed to ensure adequate proficiency levels.

CMM 8. User identification, authentication and authorization are standardized.

Policy:
Physical and electronic access to XYZ Corporation information assets must be controlled in a manner that effectively precludes the compromise of confidentiality, integrity, and availability.

CMM 9. Security certification of staff is established.

Policy:
Individuals with access to information assets belonging to XYZ Corporation and its Affiliates must undergo a background investigation, sign a confidentiality agreement, and have demonstrated proficiency and competence in their areas of responsibility.

CMM 10. Intrusion testing is a standard and formalized process leading to improvements.

Policy:
All technical service providers, whether internal or external, to XYZ Corporation and its Affiliates must construct, manage, operate, and maintain systems in a manner that ensures the availability, integrity, and confidentiality of information assets owned by XYZ Corporation and its Affiliates.

CMM 11. Cost–benefit analyses supporting the implementation of security measures are increasingly being utilized.

Policy:
Key security control objectives and their linkage to business objectives shall be formally defined and be aligned with the control documentation and requirements of ISO 27001. Controls must address defined control objectives and be implemented, tested, managed, and maintained to assure the management of risk to acceptable, defined levels, and processes must be implemented to provide continuous monitoring and relevant metrics on the effectiveness of controls, and must provide adequate warning of control failure.

CMM 12. Information security processes are coordinated with the overall organization security function.

Policy:
A steering committee comprised of senior representatives of all significant organizational departments and divisions shall be formed with a charter and scope to ensure that all assurance functions are integrated, and that risks are identified, prioritized, and managed appropriately.

CMM 13. Information security reporting is linked to business objectives.

Policy:
Information security objectives for XYZ Corporation shall be defined and a strategy developed and maintained that provides direct linkages to organizational strategies and objectives. A governance structure and framework shall be developed that describes the combination of technical, operational, management, and physical security controls in relation to the organization's technical and operational environments.

CMM 14. Responsibilities and standards for continuous service are enforced.

Policy:
Information systems infrastructure shall be managed to ensure that system configurations are in conformance with published security standards and security base lines are maintained. Change management shall be a formal process encompassing all changes capable of adversely impacting security and a summary of changes, potential risks and impacts, and applied risk mitigation measures shall be supplied to Corporate Security on a timely basis.

CMM 15. System redundancy practices, including use of high-availability components, are consistently deployed.

Policy:
All technical service providers, whether internal or external, to XYZ Corporation and its Affiliates must construct, manage, operate, and maintain systems in a manner that ensures the availability, integrity, and confidentiality of information assets owned by XYZ Corporation and its Affiliates.

12.5.1.3 Other Policies

Although the foregoing policies will serve to support the CMM objectives and provide examples of appropriate scope and detail, they will not constitute a complete set and others will need to be added. Obviously, there are various ways to categorize policies and some of the foregoing could possibly be combined as well.

At times, it may be difficult to determine what is a policy and what is a standard, especially if the policy is very detailed. A standard can be defined as something of a constant value against which something else can be measured. A policy states intent and direction; it cannot be referenced as a metric and does not address what, how, or how much. A standard can be to a greater or lesser extent exact, such as a gram, which is quite precise, whereas a standard can also provide just the outer boundaries and limits, such as an encryption standard that specifies 128 bit AES or equivalent, specifying only minimums. From a security governance perspective, it is preferable to specify the widest acceptable boundaries consistent with maintaining security in order to maximize procedural options.

An example of a typical suite of governance policies based on practical experience includes:

P 1.0 Information Security Governance Policy
P 2.0 Security Management Policy
P 3.0 Roles and Responsibilities Policy
P 4.0 Information Risk Management Policy
P 5.0 Access Control Policy
P 6.0 Information Asset Classification Policy
P 7.0 Data Management Policy
P 8.0 Incident Management and Response Policy
P 9.0 Information Asset Management Policy
P 10.0 Personnel Security Policy
P 11.0 Operations Policy
P 12.0 Outsourced Service Providers
P 13.0 Electronic Communication Policy
P 14.0 Physical Security Policy
P 15.0 Controls, Monitoring, and Metrics Policy
P 16.0 Compliance and Enforcement Policy
P 17.0 Acquisition Management and Resource Allocation Policy
P 18.0 Awareness and Training Policy
P 19.0 Infrastructure Management Policy
P 20.0 Process and Control Documents Policy
P 21.0 Communication and Reporting Policy
P 22.0 Noncompliance and Variances Policy
P 23.0 Acceptable Use Policy

Each of the sample policies in the foregoing CMM example are addressed under these 23 policy headings and it should be understood that there is no single right way to address policy construction. There are simply ways that will be more effective and better serve the organization. However, variations of the foregoing 23 policies are the culmination of several decades of experience in policy development in a number of organizations and have proven to be an effective approach. In practice, there has not been any activity that could not be reasonably classified under one of these policies, although in some instances it was the necessary to reference one or more additional policies.

The practical reasons for the broad categorization is to create a hierarchy that is readily applied and referenced based on the major security and risk management categories found in virtually all organizations. It also serves well to directly address high-level objectives as an expression of management intent and direction, and to get management acceptance with the understanding that the devil is in the details. This approach also counters the tendency in some organizations to attempt to create a specific policy for every possible eventuality, which in one major medical institution resulted in over *four thousand* policies.

12.5.2 Standards

Standards must be owned by security management. They are the primary tool for setting the measures by which policy compliance is determined and enforced. They are the primary instrument of operational governance by setting the boundaries within which people, processes, and technology must operate. Collectively, standards also set the security baselines across the organization by defining the minimum security limits.

Typically, each policy will require a number of standards primarily *divided by security domains*. The standards for low-security domains will obviously be less restrictive than those for critical high-security domains. The number of security domains will be a function of how many classifications exist for criticality and sensitivity. Most organizations will have three or four, for example, public, internal, restricted, and, perhaps, secret or in confidence. Each classification needs to be explained with as little ambiguity as possible in common language to enable personnel to classify appropriately. This is commonly done based on the impact of a compromise of a particular classification level. For example, a secret designation would apply if compromise would seriously put the organization in peril, result in danger to personnel, or cause some significant level of liability or monetary damage. A well-designed security program will subsequently address each of these classifications with protection of suitable proportionality.

12.5.2.1 Attributes of Good Standards

Standards must be carefully crafted to require only necessary and meaningful restrictions, that is, necessary in the sense of managing some credible risk within acceptable limits in a sensible way that is minimally disruptive to operations. To illustrate the point, an argument could be made that security would be significantly improved by requiring 26 mixed-character passwords to improve access control. This requirement would meet none of the aforementioned criteria. It is unlikely that a business case could be made for the requirement of 26 characters or that it would improve access control in any proportion to the disruptive effects on users. And since no one would remember such long passwords they would inevitably write them down, rendering them more likely to be compromised, and that just is not sensible.

Since a standard is a control to mitigate or manage risk, it is necessary that the control objective be known. The degree to which a standard mitigates risk is likely to be a function of how restrictive the standard is and how well it is targeted to the threat. It is not uncommon to see standards that have little or no relevance to the risk being addressed. This is wasteful and must be avoided by being clear about what needs to be controlled in order to address the risk. For example, if the risk is getting a citation for speeding, do not create standards for the color of the car.

The first attributes then will be:

- Meaningful
- Necessary
- Specific

A typical governance structure may well have a hundred or more standards in three or four security domains. It will be important to consider them collectively for consistency and continuity as well as brevity, proper construction, and clarity. Consideration must also be given to how conformance to standards will be measured and compliance achieved. Standards must be essentially binary, that is, compliance should not fall in grey areas; either the standard is met or it is not. Obviously, standards that are not achievable will not make much sense either and proportionality to the risk being addressed is necessary. Tough standards for low risk activities simply invite noncompliance.

Since standards collectively set the baseline for security in the organization, it is important to ensure that they are contiguous and do not leave exploitable gaps in protection. Consistency is required to maintain approximately the same baseline security for all assets in a particular classification. In other words, all assets classified as "critical" should have the same level of protection.

So the next attributes will be:

- Consistent
- Clear
- Contiguous
- Proportionate
- Measurable
- Enforceable

It will also be important to devise a review, modification, and acceptance procedure that includes those that must enforce the standards and those that must operate within them. This will typically include auditors and the members of a security steering committee and perhaps other stakeholders.

Exception processes must also be devised for situations in which it is not possible or not feasible to comply with the standards. Risks and potential impacts associated with exceptions must be assessed and addressed.

So we can add the attributes of

- Acceptable
- Adaptable

This is essentially a taxonomy for standards that can be used to guide and assess standards construction:

- Meaningful
- Necessary
- Specific
- Consistent
- Clear
- Contiguous
- Proportionate
- Measurable
- Enforceable
- Acceptable
- Adaptable

12.5.2.2 Sample Standards

Developing the standards for multiple security domains for two dozen or so policies can be complex and arduous. In the following section, we will create some sample standards based on work with commercial organizations during the past decade.

One practical admonition from experienced security managers is to understand that auditors will hold the security manager to what is written in policies and standards and it will be prudent to ensure some "wiggle room" where needed.

For our sample standards development, let us consider a critical, all-encompassing policy regarding access control. Our deceptively simple access control policy is:

> Physical and electronic access to XYZ Corporation information assets must be controlled in a manner that effectively precludes the compromise of confidentiality, integrity, and availability.

Considerable work could be required to develop a suitable set of standards that meet all of the aforementioned criteria in terms of providing the necessary controls without being needlessly restrictive.

First, we must consider how many classification levels must be addressed, with consideration of the system and infrastructure capabilities and constraints. That is, setting standards that cannot be generally met will obviously create a problem. Some legacy systems might only support six-character passwords, so setting standards that require eight will create instant exceptions that must be considered.

12.5.2.3 Classifications

For most organizations, three or four security classifications will usually be adequate and may be excessive. These can include internal, confidential, and restricted in addition to unclassified or public. The granularity of a greater number of classifications may not be warranted by the additional implementation and management requirements. Whatever the decision on the classification levels, optimally, standards will be required for access control for each level, both physical and logical. This is to avoid excessive restrictions on low-criticality assets and to maintain proportionality.

What access must to be considered? Given the aforementioned general access control policy, logical access to applications and systems as well as physical access to facilities must be addressed. Logical access will usually be handled by user ID and passwords so we will need a standard for identification and password authentication. The following is a sample from a project to develop tightly controlled standards for a financial organization, which may be excessive for many organizations but serves to illustrate possible approaches.

Standard Statement

Passwords are the primary authentication method utilized to control access to XYZ Corporation. IT systems and resources. This standard sets forth the minimum criteria necessary to ensure adequate security for these systems and minimize the risk of compromise. In addition to the standard, it is the responsibility of all personnel to maintain the security and confidentiality of passwords and to report to Security any situation that may result in compromise.

This standard must be used when passwords or personal identification numbers (PINs) are used as an authentication mechanism. The standard provides requirements for the following:

- User communities
- Password/PIN format
- Password usage
- Password expiration
- Password control function
- Failed log-in attempts
- Password change process
- Password reset
- Password reuse
- Password generators

- Password storage
- Event logging
- Password transport
- Inactivity timeout

User Communities

There are several user communities or types of passwords used within XYZ Corporation. These communities are based on the type of entity being authenticated:

- Employee user ID
- Administrative and superuser ID
- Contractor user ID
- Business partner user ID
- Customer user ID
- Service user ID

Password Format

The password format (complexity and length) determines the relative strength of a password. Passwords must conform to the minimum following format constraints, dependent on asset classification:

- Restricted: Minimum length eight characters
- Confidential: Minimum length ten characters
- Secret: Minimum length 12 characters
- Maximum length 16 characters
- Mix of both alpha and numeric characters
- At least one nonalphanumeric character
- At least one upper-case character
- The numeric value cannot exist exclusively in the last position of the password (i.e., if the password contains only one numeric value, this numeric value may not be in the last position of the password).
- May not be the same as the user ID
- Identical characters may not be adjacent

PIN Format

PINs must conform to the following requirements:

- Minimum length five numbers
- Maximum length 12 numbers

Password Usage

Passwords must be entered by means of individual keystrokes, not scripted in a function key, Web page form, or hot button, and entry must not be displayed on the monitor.

Password Expiration

All passwords must expire. Users must be notified, when possible, two weeks prior to password expiration. Different communities of passwords may expire at different intervals, not to exceed the following maximums:

Community of User ID	Password Validity Period
Employee User ID	180 Days
Administrative/superuser ID	90 Days
Contractor user ID	180 Days or length of contract, whichever is less
Business partner user ID	180 Days
Customer user ID	365 Days
Service user ID	365 Days

Password Control Function

When passwords are used as the user authentication scheme for a system, the system must provide the following functions:

- Password expire function
- Password expiration notification function
- Password change function
- Password format enforcement function
- Failed log-in enforcement

Failed Log-in Enforcement

The authentication system must be configured so that the system will accept three attempts within a 24-hour period, then disable the associated user ID for a period of 10 minutes. Alternatively, the system may be configured so that the help desk must be contacted to reset the password subsequent to identification and authentication by two specific verifiable items.

Password Change Process

The system password change process must support:

- Reauthentication as part of the password change process
- All format and standards shall be checked and enforced.
- New passwords will be entered twice to ensure no errors.
- Users must be notified by e-mail that the password change was successful.
- A password must not be changed more then three times within a 24-hour period.

Password Resets

When passwords are reset by a third party or a remote service, the following must be supported:

- Users must be authenticated prior to any system or password resets, using a combination of two verification or attestation methods.
- Users must be informed of the new password through an approved secure means.
- User must change password upon first use.
- Temporary passwords must expire within 72 hours.

Password Reuse

Systems shall not allow a password to be reused for 12 months or 15 passwords, whichever is least.

Password Generators

Random-password generators are permitted and desirable when humans are not the users of the password. Password generators must comply with this standard when used.

Password Storage

- Passwords are classified at the highest level.
- Systems being used to store passwords must be hardened.
- Passwords must be stored in an encrypted format using an approved process (i.e., approved one-way hash algorithm, such as SHA-1, MD5).
- Only authorized processes/systems may replace passwords within a data store.
- Passwords must not be "hard coded" into files, scripts, or programs.

Backup and Recovery

- Password stores may be backed up as part of the business process, provided the backup medium is classified and treated at the highest level.

- Password files must not be restored (passwords will be reset instead of restoring old passwords).

Event Logging

All applications using password authentication must log the following events and time of occurrence:

- User log-on
- User log-off
- User log-off reason, for example, user timed out or user logged off

Systems providing password storage and management must log the following events and time of occurrence:

- New user created and by whom
- Password change by user
- Password reset by whom
- User deleted and by whom
- For which application or system (if needed or performed remotely)

Password Transport

Passwords shall not be transported or stored in clear text form.

Inactivity Timeout

When an inactivity timeout mechanism is used to secure a system from abandonment, the period of time shall not be greater than 20 minutes. The user must reauthenticate to obtain access. The reauthentication process may not be abbreviated in any way.

Scope

This standard addresses the treatment and characteristics of passwords and PINs used by XYX Corporation at the enterprise level. The standard applies to all information technology systems both external and internal facing, regardless of geographic location. Passwords for employees and agents of XYZ Corporation and its Affiliates must meet the criteria set by this standard.

Consequence of Noncompliance

Failure to comply with this standard may result in corrective action up to and including termination and/or civil or criminal prosecution. XYZ Corporation reserves the right to disclose to anyone, at its discretion, any information at any time without limitation.

Responsibility for Compliance

Any employee, individual, or entity with access to information assets belonging to XYZ Corporation and/or its Affiliates is responsible for complying with this information security standard.

Management of each department within XYZ Corporation and/or its Affiliates is responsible for ensuring that all employees understand and adhere to all information security policies, standards, and procedures. Management of each department within XYZ Corporation and/or its Affiliates is also responsible for noting any noncompliance, informing CIS and HR, and initiating corrective actions.

Rationale

Passwords represent the first line of defense for virtually all information assets. Sharing of passwords, weak passwords, and infrequently changed passwords all pose easy targets for unauthorized users.

The foregoing standard makes a distinction for authentication requirements between various security classification levels. In some situations, stronger authentication may be required for highly sensitive or critical assets and two-factor implementations may be warranted. In some situations, an organization may utilize the authorization standard to define access control for the various classification levels.

12.5.2.2 Physical Access Standards

Another standard will be required for physical access controls to facilities with various classifications at the same level of detail as the logical access standard above. Coordination of these standards is essential and a number of specific requirements must be specified in the appropriate standards. For example, high-criticality information systems cannot be located in low-security facitilities; with mixed classification assets, the environment and access must be classified at the highest level.

A suite of access control standards might include:

- Entitlements or resource access control system
- Authorization standard
- Implementation standard
- Identification standards
- Authentication standards
- Access standards
- Visitor access
- Access to customer data

- Privileged access
- Developers access
- Customer restrictions
- Physical access to controlled areas
- Auditability

The standards might include provisioning and onboarding requirements as well. There will be media issues such as wireless access, external VPNs, trading partners, and, perhaps, some customer access to parts of the system. There will be physical as well as technical and logical matters to consider and there can be concerns about how to handle access from employees working remotely. It may also be necessary to consider access control monitoring, operational status, requirements for warnings and notifications of control failure, and any other required metrics, entitlements, and, perhaps, other control measures.

12.6 SUMMARY

The creation of complete standards can be a daunting task, although considerable assistance can be provided by standards such as CobiT, ISO 27001 and 27002, as well as NIST 800 series publications and others. In many instances, a significant number of standards will exist in the technology realm and the majority of development efforts will be on nontechnical procedural activities. In any event, since the standards collectively set the security baselines for an organization, it is critical that this process be thorough and complete. Experience suggests that an overhaul of standards in a typical organization can take a year, or considerably more when including ratification and implementation.

Chapter 13

Security Program Development Metrics

Information security program development is the process of implementing strategy into operational use, including the structures and processes of governance. This will typically result in a series of projects or initiatives as the result of developing a security strategy that meets the strategic objectives of the organization. It is the series of activities that over time results in achieving the desired state of the information security program, as discussed in Chapter 7.

The projects undertaken to achieve program objectives can utilize standard project management approaches such as Prince II and metrics not unique to information security. This will typically include GANNT, PERT, and critical-path charts. Assuming that preconditions such as specific objectives have been defined, adequate resources committed, and a road map or architecture developed, the relevant question is what decisions must be made and what metrics will be needed to provide the information required to manage security program development?

This will, in large part depend on the scale and nature of security program implementation. If it is a major initiative and highly complex with many moving parts, such as implementing an enterprise resource planning system (ERP), there will be many dependencies and serious consequences for poor planning, inadequate metrics, and ineffective management. Whatever the scale, adequate monitoring and metrics becomes a crucial element in managing for satisfactory outcomes.

13.1 INFORMATION SECURITY PROGRAM DEVELOPMENT METRICS

From a program development perspective, what information will a security manager need to make implementation decisions and what metrics will provide the required information? Numerous minor operational decisions will typically relate to resolving particular problems encountered during implementation. The more important management decisions will relate to whether implementation is meeting budget and time objectives, and, ultimately, whether the project performs as anticipated. Time and budget issues can be crucial if they are in the critical path of other initiatives. Decisions may need to be made whether to increase or reallocate resources or perhaps reschedule dependent activities. These aspects can be very important if major organizational activities are dependent on timely completion of a security initiative and its meeting the required performance objectives. An example might be an online business such as banking or a product merchandising initiative dependent on an adequate and functional security implementation or achieving regulatory compliance mandates within an allotted time frame.

Key goal indicators and key performance indicators are likely to be the most useful approaches to measuring program development progress. Critical success factors (CSFs) should also be determined during the planning phase.

A formal process such as the system development life cycle (SDLC) approach should be used to define phases and ensure completeness. There several variations on the process but they all generally include:

- Determining feasibility
- Establishing requirements
- Solution architecture and design
- Proof of concept
- Full development
- Integration testing
- Deployment
- Quality and acceptance testing
- Maintenance
- End-of-life decommissioning

If the SDLC approach is used in security program development, each phase should include KGIs and KPIs as metrics, and CSFs should also be identified.

For example, a feasibility study goal indicator will be the completion of an analysis determining whether or not an initiative is likely to meet the defined business objectives at acceptable costs and risks within required time frames. Performance indicators for this phase will be whether the study provides answers with adequate certainty, progresses at the desired pace, and is completed on time at anticipated costs.

Establishing requirements goals could include ensuring that all critical business needs can be addressed. Performance will revolve around whether necessary specifications are achievable, affordable, and can meet requirements.

Solution architecture and design goals will relate to completion times, incorporating all required elements, design efficiency, meeting control objectives, and so on.

Proof-of-concept goals will be to prove assumptions, feasibility, and designs, and provide evidence that the required performance objectives can be achieved.

Full development and coding will be a project to be managed in conventional ways with standard project metrics in terms of milestones, costs, resources and so on, as will integration testing.

Deployment objectives and performance goals will also include time, cost, and effectiveness measures.

Quality and acceptance testing (QAT) will include metrics on the outcomes and deficiencies encountered as well as conformance to specifications.

Maintenance metrics will be gathered over time and the performance indicators will be set by the original specifications.

End-of-life decommissioning may include certified destruction of media and other assurance measures.

Although this may seem to be a highly granular approach, for important and/or complex projects, it is prudent to set goals and performance indicators for each phase to manage project risks and ensure that development proceeds at the necessary pace.

13.2 PROGRAM DEVELOPMENT OPERATIONAL METRICS

Operational metrics for program development will center on specific project-related issues. Typically, decisions required will deal with project progress, meeting milestones, unforeseen impediments, inadequate performance, and so on. The information needed will in like fashion be about milestones and project performance-related issues. They will deal with design problems and implementation issues. The goals at this level will hopefully be clear and in terms of SDLC, the issues will be far down on the list of requirements for projects, that is, objectives, feasibility, and design elements will already be in place.

Chapter **14**

Information Security Management Metrics

Metrics only serve one purpose—decision support. We measure to manage. We manage to meet objectives in order to achieve desired outcomes.

Strategic measures and metrics for governance were discussed in Chapter 5 and are basically the navigation tools to keep a program on track.

Management metrics, discussed here, are concerned with the effectiveness of a program and providing the information needed to fly straight and level and operating within acceptable limits. These are tactical metrics that are needed to keep the program operating at acceptable levels guided by the strategic objectives—the destination. Just as operational metrics can be "rolled up" to provide assurance to management that the machinery is performing within acceptable ranges, so too can management metrics be abstracted to provide assurance to executive management that the entire security program is operating within proscribed limits so that strategic objectives will not be compromised.

Although technical metrics for information security have seen significant gains in the operational area, strategic and management measures remain virtually nonexistent. Consequently, there is scant information to provide the basis for the decisions needed to manage an information security program on an ongoing basis.

Some practitioners contend that abstractions of technical operational information can be effective for security management but this argument is flawed. There is simply no way of putting technical operational information together to navigate or manage a security program. It is not unlike attempting to operate an aircraft by using fuel, oil pressure, and engine temperature gauges. From a predictive perspective, the only thing known from operational information is that objectives cannot be achieved if operations fail.

As discussed in Chapter 5, the manager of the aircraft, the pilot, requires three types of information in order to accomplish his or her responsibilities.

1. Strategic information, which is provided by navigation systems
2. Management, or tactical, information such as attitude, airspeed, turn, and bank
3. Operational information including oil pressure, temperature, and fuel level

Navigation information is strategic for the flight as it tells the pilot in which direction to go and how far it is to the destination, as well as the current location. The destination is set by senior management based on the strategic requirements and operation of the business, and can be considered the business objectives of the flight needed to operate an airline.

The pilot needs management information regarding the aircraft attitude, altitude, airspeed, direction of flight, and so on to operate it safely within acceptable limits and head it toward the destination set by senior management. It will also be important to keep an eye on operational metrics such as fuel, oil pressure, temperatures, and so forth to ensure that they are within acceptable ranges, which are analogous to technical IT metrics, to ensure that progress is not disrupted by mechanical failure.

The requirements for a security manager operating a security department are not much different.

The information needed will be:

Strategic (to determine direction)
- Objectives (destination)

Management (to determine attitude and heading)
- Current state (location)
- Management status (operating within acceptable limits)

Operational (to determine system functionality and impending issues)
- Status of the machinery (operating in the green)
- Performance trends (predictive of issues)

14.1 MANAGEMENT METRICS

Management metrics differ from strategic governance and program development metrics in several ways. Just as the metrics for running an airline or building an airplane are quite different from the ones used to fly it, so too are the metrics needed for day-to-day management and administration of a security program.

14.2 SECURITY MANAGEMENT DECISION SUPPORT METRICS

The information needed for management of a security department is also quite different from the information needed for operational security management. Operational security is concerned with information about such things as server configurations, intrusion detection systems, firewalls, and so on.

For managing a corporate security department, let us consider the types of decisions that are typically made by someone serving in the CISO role, regardless of the actual title. This role should generally report to senior management or the audit committee of the board of directors. The CISO will usually be involved in strategic activities such as developing policies and standards, incident management, developing or implementing security strategy, budgets, personnel issues, and regulatory and contractual matters.

As a part of an oversight function, a CISO may also monitor some operational information, perhaps in summary form, to ensure that the "machinery" is performing within normal limits, just as a pilot wants to know that the engines are operating normally and there is adequate fuel. This information is important only from the aspect that "mechanical" failure can cause operational failure, resulting in objectives not being realized. But this operational information tells the pilot nothing about whether the aircraft is headed in the right direction to arrive at the desired destination.

Management of information security will be primarily concerned with strategic business outcomes, ensuring that security activities properly support the organization's objectives and alignment with the overall business strategy. As a result, the types of information needed must be indicative of things like the effectiveness of risk management, the appropriateness of resource allocation and utilization, and the degree of overall policy compliance. If we accept the notion that decisions can only be as good as the information they are based on, then it follows that, for most organizations, management of security is far from optimal, as evidenced by the ongoing security failures reported in the press.

The starting point for determining the kinds of measures, metrics, and monitoring needed is to understand the scope and responsibilities of those that make the decisions. This provides the basis for determining the information required, which in turn defines the kinds of measures and monitoring processes that must be developed. The areas that are generally the focus for most security managers will include the aforementioned six outcomes:

1. **Strategic alignment**—aligning security activities in support of organizational objectives
2. **Risk management**—executing appropriate measures to manage risks and potential impacts to an acceptable level
3. **Assurance process integration/convergence**—integrating all relevant assurance processes to improve overall security and efficiency
4. **Value delivery**—optimizing investments in support of the organization's objectives
5. **Resource management**—using organizational resources efficiently and effectively
6. **Performance measurement**—monitoring and reporting on security processes to ensure that organizational objectives are achieved

In previous chapters, key goal indicators and key performance indicators were discussed in the context of security program development. For ongoing information security management, performance measures are generally not as relevant. Rather, information about location, direction, and speed will be more useful in guiding the overall security program. In other words, performance measures, though useful for operational activities, do not provide navigational information. Knowing how well the compass is working will not tell you the direction in which you going.

14.3 CISO DECISIONS

Effective security management currently relies primarily on the experience of the security manager coupled with a shotgun approach to plugging all perceived vulnerabilities to the extent possible and employing "best practices." Given a modicum of luck and adequate resources, this can be reasonably effective much of the time. However, as witnessed by the near continuous reports of increasingly costly compromises, this approach fails all too often. Typically, these failures are not a surprise to security managers and are the result of organizations failing to support and implement recommended security measures.

What is considered an adequate level of security is often an area of contention between IT, business owners, and security managers. In part, this is a consequence of security managers failing to make a persuasive business case for controls seen as too restrictive or costly. It is also in part due to the fact that, lacking effective metrics, there is a tendency for cautious security managers to adopt excessively restrictive controls, much as the lack of an accurate speedometer will cause a cautious individual to drive too slowly. The uncertainty caused by a lack of adequate metrics drives prudent security managers to attempt to increase safety margins, which is often seen as overkill by those focused on performance.

Effective information security management metrics, then, are those that provide the right kind of information for the security manager to make appropriate decisions. The key to developing these metrics is *defined objectives,* which provide the reference points against which to measure.

If the previously stated six outcomes for information security are to be achieved, it will be necessary to provide clear objectives for each of them. This will, in turn, provide the reference points by which information security management metrics can be developed. The decisions that a CISO must make are then determined by whether security activities are headed in the right direction to achieve those objectives.

Although the six outcomes are readily understood, reducing them to concise objectives requires some effort. Achieving consensus on the extent and degree that these outcomes should be attained will provide the basis for the objectives needed to design useful measures, metrics, and monitoring activities.

14.3.1 Strategic Alignment—Aligning Security Activities in Support of Organizational Objectives

The concept of aligning security to support the organization's objectives is readily understood but the questions raised are more complex. What degree of alignment is adequate? How can it be measured?

More alignment is arguably better than less, so the metric should not be binary, that is, aligned or not aligned. The metric will also be subjective and not quantitative. Nevertheless, the extent to which security activities are *perceived as hindering any particular organizational activity* is likely to be a useful measure, whether true or not. But the metrics for strategic alignment must extend beyond just perception and deal with reasonably quantifiable measures to manage those perceptions. To accomplish this, it will be necessary to determine:

1. Clearly defined objectives of the organization
2. The risks and security implications for those objectives
3. The boundaries of acceptable risk

For example, if one of the objectives of the organization is to provide online banking, the risks and implications can be assessed and potential impacts of compromise determined. Management must then determine the extent to which various risks associated with the activity are acceptable or not. Alignment of security with this objective is farily straightforward, although likely to result in some tension between usability by customers and adequate safety of the process. The perception of alignment is likely to be the result of how great an impediment to implementation security considerations are seen to have.

Another example of a more tactical nature is a scenario in which a security manager determines that eliminating e-mail attachments will significantly reduce virus infections of the organization's systems. From a purely safety perspective, this might be prudent but from a business perspective is likely to adversely impact operational efficiency and be perceived as an unnecessary nuisance and contrary to the notion of strategic alignment.

What information is needed to make this decision? To a large extent, this will be determined by the industry sector. It is common in manufacturing and retail to preclude any security activity that has a noticeable negative impact on operations. In this case, the information needed by the security manager is simple; for example, the metric can be the number of complaints from users. Any increase in complaints mandates the requirement to make controls less restrictive.

The regulated financial sector is more balanced in considering the trade-offs between performance and safety and might require the development of a business case assessing the range of options and the costs and benefits. The information needed to support a decision in this regard includes:

1. Costs of infections from e-mail attachments
2. Value of benefits to users and/or cost of inconvenience
3. Cost of incremental limitations
4. Costs of remedial processes as a result of attachments

The metrics for the cost of infections are relatively straightforward. There is no consistent methods used across industries but any reasonable approach should be satisfactory as long as the methods used are consistent. This will allow trends to be determined, which might be a basis for different decisions.

Metrics for inconvenience and incremental restrictions can be measured against a baseline of user complaints or the results of periodic surveys. Incident costs resulting from infectious attachments can be based on information from incident reports and response and resolution costs from postmortems.

Decisions made around strategic alignment regarding appropriate risk baselines, acceptable impacts on the organization from security activities, and acceptable trade-offs between performance and safety pose more of a challenge, in part because these will often be dynamic elements and in part because it will be difficult to achieve consensus among stakeholders with very different perspectives and agendas. Often, this results in highly subjective or reactive decisions and a focus solely on quarterly financial performance. The focus on profit regardless of risk is a typical result of misaligned incentives and inadequate governance.

Case Study

Consider the approach undertaken by a major bank to align security with its business objectives. The organization was clear that it wanted to expand its operations online because the efficiencies gained from online banking had proven persuasive. It was also demonstrated by surveys and analysis undertaken by the CISO that customer trust was the single most important factor in driving growth and the factors that most affected that trust were primarily in the information security domain. In response to these findings, senior management implemented a plan whereby several hundred of the top managers have a substantial portion of their compensation tied to evaluation of their adherence to security requirements and compliance of their respective organizations.

The revelation in this instance is that the most significant security metric had nothing to do with security; rather, it was customer satisfaction. In the case of this bank, careful analysis made it clear that this was the best correlation with growth and inadequate security was one of the main causes of unhappy customers.

The important lesson is that for many organizations, at the strategic and management levels, the best security metrics may only be related to the impact of security, not any measure of security itself. Decisions related to strategic alignment will often be strategic as well.

For example, the decision to make standards more restrictive to increase security baselines over time may affect the entire organization and is likely to be costly. What information is needed by the CISO to make that decision and be able to support it?

1. Evidence in the form of studies or surveys that security compromise would result in a loss of business. An example is the PGP/Vontu analysis of a number of breach losses showing that 19% of customers immediately ceased doing business with the affected organizations and another 40% were considering doing so.
2. Based on the evidence, determine probable lost revenues and profits.

3. Perform an analysis to estimate the decrease in security failures based on historical evidence of preventable breaches.
4. Determine costs of modifying and implementing new standards, the modification of procedures, and new technical control measures.

Analysis of this information can be made using ROSI or other means and will determine the optimal level of security baselines based on costs and benefits. If it is determined that increasing security baselines makes business sense, other measures will be required to measure progress of the initiative as well as the ultimate success in achieving the objectives. These could include project management metrics of implementation against plan and budget. Subsequent surveys could assess customer satisfaction and the correlation with sales or, perhaps, a decrease in customer turnover.

14.3.2 Risk Management—Executing Appropriate Measures to Manage Risks and Potential Impacts to an Acceptable Level

To devise effective metrics for risk management, it must be determined what constitutes "appropriate" risk management measures and what is considered an "acceptable" risk level, Without fairly concise definitions, it is difficult to make decisions about what levels risk must be managed to and in what direction the program should be headed. Wide variances in these elements between different organizations, sectors, and cultures will affect what is considered appropriate and acceptable.

The organization's *risk tolerance* is the major determinant in what is considered acceptable risk and the prudent security manager will determine this level by some measure. Determination of acceptable risk levels is often easier to achieve by discussing impact levels in terms of the range of likely costs of disruptions and the probability of occurrence. If, for example, management decides that any event costing less than ten thousand dollars that will occur less than once a year is an acceptable impact not worth mitigation efforts, the security manager has a defined target and a reference point for a useful metric.

The extent to which the culture of the organization is risk adverse is a major factor in what is ultimately determined as "acceptable." This, in turn, will likely be the main determinant of the level of support for risk management activities. Although it is possible in an organization with a high aversion to risk to require an excessive level of risk management, resulting in poor cost effectiveness, the more common problem is just the opposite. The undermanagement of risk coupled with good fortune, resulting in a lack of incidents, is typically a greater problem because it may suggest to management that existing risk management efforts are adequate and, as a result, they might opt to allocate resources to areas perceived as more pressing or offering greater returns.

For most organizations, risk management for information security is generally ad hoc and haphazard. This is born out by studies showing that security resource allocation is generally unrelated to risks or impact and is often narrowly focused only on IT risks. The lack of a mature security governance structure with the resultant absence of defined objectives is often the cause and the result is that some typical IT risks are managed reasonably well whereas others are not addressed until there is a compromise. Analysis of breach losses by PGP/Vontu in 2006 highlights this situation (Table 14.1).

Although the usual emphasis on purely preventive IT perimeter controls appear to have been generally effective at keeping intruders out, it can be inferred that baseline security of the organizations studied were highly inconsistent and most non-IT risks were not managed effectively. Consistent security baselines would arguably result in roughly equal losses from all causes, of course, within defined, acceptable levels.

14.3.3 Metrics for Risk Management

It is clear that a requirement for developing effective metrics for risk management as for all other aspects of information security is to define and develop clear objectives to serve as the reference point against which to measure. From a risk management standpoint, the question then arises, what information will the security manager require to make the decisions to effectively guide risk management activities? The answer might be:

1. Determination of the organization's risk tolerance
2. Comprehensive resource valuation
3. Complete risk assessment
4. Business impact assessment of important systems
5. Tests of control effectiveness and reliability
6. Known level of metric accuracy and reliability

The information from metrics around each of these six items will provide the basis for determining both the priority of risk management efforts as well as the level of effort required to achieve the objective of acceptable risk and impact levels.

In the next section, we well examine what types of metrics, measures, or monitoring can be used to provide this information.

Table 14.1. Breach losses from compromise of 33 companies*

Lost laptops	35%
Third party or outsourced	21%
Electronic backup	19%
Paper records	9%
Malicious insider or malware	9%
Hacking	7%

*2006 Annual Study: Cost of a Data Breach, PGP/Vontu.

14.3.3.1 Organizational Risk Tolerance

The level of risk or impact management considers acceptable must be determined as a prerequisite to an effective risk management effort. Impact in financial terms is usually the best way to arrive at this determination as organizations operate on numbers. It should be considered that there is a risk associated with any activity as well as with not doing something. Although risk management is usually considered in terms of preventing loss from adverse events, in some cases it may also be concerned with not achieving some potential gain.

For any organization, there is some level of financial impact small enough that no expenditure to address the risk is warranted. There will also be some level of potential loss that will not be acceptable, dependent to some extent on the probability of occurence. If the probability of even a major event is vanishingly small, the risk may still be accepted. It should be noted that acceptable risk may differ for various parts of an organization and it may change over time under different circumstances, consequently requiring periodic revisits.

The determination for acceptable information security impacts is not significantly different from determining suitable levels of insurance. In both cases, less impact is generally more costly and, consequently, the exercise boils down to a cost–benefit analysis. The rational point to arrive at is where the increasing cost of mitigation crosses the decreasing cost of impacts.

Depending on the organization, there may be a number of reasons that the determination of risk tolerance may not be entirely rational, however. These reasons can range from "it hasn't happened here and we'll deal with it when it does" to skepticism about the accuracy of risk assessments or the level of threat. There may also be financial constraints that preclude appropriate mitigation activities such as conserving funds to improve quarterly results.

In some cases, it may be more effective to go through the process of developing recovery time objectives (RTOs) for critical organizational activities. The process of negotiating acceptable RTOs at an acceptable cost provides a direct financial value of what the organization is willing to spend to address the risk of system failure. This serves as a clear indicator of risk tolerance based on acceptable impacts. This approach is, of course, also dependent on the credibility of the prerequisite risk assessment in terms of the probability of an event and its potential to cause system failure.

Defining risk tolerance in measurable terms provides the objectives the security manager can manage to as well as the reference point for metrics needed to guide risk management efforts. This can also be useful in that risks that fall below this measure can provide the basis for reducing controls and costs, whereas risks in excess of the measure will require additional mitigation efforts.

14.3.3.2 Resource Valuation

An obvious requirement, though surprisingly often not the case in many organizations, is that all information resources be known. For risk to be managed appropriately, it is essential that the actual or relative value of information assets be determined. For assets of little value, there is obviously not a significant reason to manage risks to them.

The underlying principle of proportionality requires that the cost of protection never exceed the value of the asset and that assets of equal value should have the same level of protection.

Many information assets will be difficult to value with any precision and some organizations use a simpler process of just three or five levels of relative value. Though inexact, this approach is usually sufficient to determine the necessary levels of risk management effort as well as the priority.

14.3.3.3 Comprehensive Risk Assessment

Risks that are not known are not likely to be managed. Effective risk management is not possible without the information provided by comprehensive risk assessment and analysis. Comprehensive means that the entire business process is assessed from initial entry into the organization until the final output. The assessment must include all related physical processes in addition to technical ones.

In addition, external factors such as environmental, cultural, legal, and geopolitical risks, to name just a few, must be considered as well, insofar as they may pose a viable risk that may impact the organization.

14.3.3.4 Business Impact Assessment

Potential impacts from compromise must be understood in order to manage risk. Impact is the bottom line for risk management, which might more accurately be called impact management. Enormous risks that have little or no impact do not need to be managed. It must be acknowledged that there continues to be disagreement among security practitioners about the exact meanings of the terms risk, impact, exposure, and so on. For the purposes of this book, the definition of risk is the probability of a threat exploiting a vulnerability and causing an impact. The level of risk will be a function of both frequency and magnitude.

The information obtained from business impact assessments will be critical for effective risk management decisions in terms of the level of mitigation required as well a prioritizing security resource allocations.

14.3.3.5 Control Effectiveness and Reliability

Control effectiveness and reliability must be determined to consistently manage risks to acceptable levels. The degree of effectiveness will determine to what extent layering of controls is necessary to achieve acceptable mitigation levels. The information security manager can also use this information to support decisions to replace or modify controls that are not sufficiently effective or reliable. Technical controls usually fare much better than procedural controls in this regard. Metrics on effectiveness can be obtained by periodic testing or some process for regular or continuous monitoring.

Metrics on incident detection, response times, impacts, and effectiveness of response activities will also provide useful information to provide a basis for risk management decisions.

14.3.3.6 Metrics Accuracy

Metrics developed to provide information for risk management must be tested for accuracy if they are to be useful. Inaccurate or misleading metrics may be worse than no metrics. The accuracy need not be extreme, provided the limits of the range are known and can be factored into decision making processes. Other attributes of metrics are discussed in Chapter 7.

14.3.4 Assurance Process Integration

Achieving a level of integration of the various assurance functions in a large organization can be a daunting task despite the obvious rationale for doing so as previously discussed. Responsibility for the overall safety of the enterprise falls to a number of different parts of the organization that are to a large extent, interdependent; for example, logical security is not possible without physical security and both rely on HR to screen risk personnel, etc. There are several approaches the information security manager might utilize to measure the relative level of integration and to support the case for increasing it if found deficient.

Measures or indicators of the need for better integration could include:

- Incidents traceable to a lack of integration. The Australian Customs office example where two people appearing to be service personnel simply entered a supposedly high security area and walked out with several highly sensitive servers would certainly qualify as an indicator. Most organizations would hopefully consider something less dramatic as a suitable indicator of the need to coordinate and integrate these functions.
- One indicator could be that there are significantly different levels in the organizational structure of physical and information security and other major regulatory functions such as risk management, compliance, QA, etc. For example, it is not unusual for a CSO in charge of physical security to be at the vice president level whereas the information security officer is far down the corporate ladder. The obvious inference is that the organization considers information security less critical than physical security.
- Inconsistencies or contradictions in the objectives, policies and standards applied to various assurance functions.
- No point of common reporting for assurance processes in the organizational structure.
- No references to functional interfaces between assurance processes in roles and responsibilities.
- An absence of communications between assurance providers.

If these indicators are sufficient to make the decision that an increase in assurance process integration is needed, they will also be suggestive of how that may be accomplished. This can include:

- Creating a policy to mandate integration supported by standards for minimum requirements and procedures for how that will be accomplished.
- Raising the issues at a common juncture in the organizational structure where it can be addressed such as an executive committee.
- Modifying roles and responsibilities to explicitly require interfacing with other assurance functions
- Mandating the regular meeting of high-level representatives of the relevant departments with a charter to provide continuity of assurance functions.

Measures of the level of assurance integration can be approached in several ways. For example:

- A steering group comprised of senior representatives from each of the groups mentioned above that meets on a periodic basis would be a good indication that there is recognition that their collective efforts provide overall information assurance; that their efforts must be contiguous and coordinated; and that their will be assurance synergies possible. Measures could include the extent that assurance integration is the topic of discussion and specific actions taken by the group to improve communication and coordination.
- A simple but potentially useful indicator of the level of cooperation and communication between these functions could be the level of e-mail interchange between these parts of the organization. It is reasonable to assume that greater communication will increase the level of understanding and integration.
- A standard security review of the relevant functions to see where interfaces exist and when and under what circumstances there is information and responsibilities exchanged.

14.3.5 Value Delivery—Optimizing Investments in Support of the Organization's Objectives

What information will a CISO need to guide decisions on an optimal investment strategy for security? There are many possible elements to consider. The first step may be to develop measures or metrics to determine whether existing security controls are adequate and cost-effective.

As in any other activity, the degree of standardization can be a useful indicator of value insofar as multiple solutions to the same problems will be more costly. For example, standardized access controls will be easier to manage and monitor than if they are all different in various parts of the organization. From a security perspective, homogenous controls, though easier to manage, will also add a dimension of risk in that a common vulnerability will aggregate risk. The benefits of standardization will, therefore, to some extent be offset by the need for greater robustness to offset the increased risk associated with a common failure mode.

It may be warranted to attempt to determine financial return on investment of new or existing security investments to provide the information needed for good decisions. Various efforts to calculate return on investment in security have mixed results. Some view security more as an insurance policy and maintain that trying to calculate a return is not productive. Yet it must be recognized that virtually all organizational activities are guided by some form of cost–benefit analysis and, ultimately, security activities must be as well.

Some types of security investments readily lend themselves to financial analysis, whereas others pose more of a challenge. The cost savings for automating certain activities can be easily calculated. The financial benefits of preventing certain events will be more difficult to do with any supportable accuracy. It may be the problem of proving a negative but there have been efforts that may be helpful. ROSI proposes to assess return on security investments by the amount of the reduction of losses in relationship to the investment. The formula is discussed and shown in Chapter 6, Section 6.1.4.

The reduction of losses will often be speculative but in some cases may be supportable or even accurate. Historical data may show fairly consistent levels of losses over time that a specific course of action can reduce to a definable extent. The decisions will again rest on the most cost-effective means of achieving security objectives. The information needed will be:

- Objectives
- A measure of the required effectiveness
- The degree of required effectiveness achieved
- Cost

High-level objectives have been discussed in general terms in Chapter 5, Section 5.1.1 but in this case are likely to be more tactical and specific. They are likely to be related to a defined control objective and the issue will be to determine the cost of the options available at comparable levels of adequate effectiveness. If the control objectives have been defined, the main difficulty will be to determine the degree of required effectiveness and to what extent a particular solution meets the requirement. The criticality or sensitivity of the resources giving rise to the control objectives must be assessed for the level of effectiveness required to be determined.

For high criticality, either a high level of effectiveness coupled with a high level of reliability will be needed or some combination of multiple controls must be layered for a "belt and suspenders" approach that in the aggregate provides an acceptable level of performance.

The decision that needs to be made will require information on:

1. The criticality or sensitivity of the resources. This can be in relative terms and approximated as low, medium, or high.
2. The control objectives in specific terms. What will the control accomplish and how will that be measured? For example, access control that will provide a 99.9% certainty that unauthorized access will not occur.

3. Will the controls considered provide the required level of certainty? Determining this may require testing and/or statistical analysis. For example, two sequential controls with a 97% certainty of precluding unauthorized access and layered should provide the required 99.9% level of certainty.
4. The issue will then be, what combination of controls will provide this level of control at what cost? The costs will need to based on full life cycle TCO computations including acquisition, deployment, operation, maintenance, testing, and so on.
5. Consideration must also be given to the fact that 99.9% certainty means that on average, for each thousand events, an unauthorized individual may gain access and the potential impact of that must be considered. Layered access controls will typically include authorization controls as well, which in combination will raise this possibility to a more secure level.

14.3.6 Resource Management—Using Organizational Resources Efficiently and Effectively

What measures are available for the information security manager to determine effective and efficient use of security resources? Considering the stated requirement, *effectively* would mean that defined objectives for security are achieved; *efficiently* would require that those objectives were achieved at the lowest cost in time and money.

The set of measures of effectiveness relates to the prior notion of value insofar as cost-effectiveness. However, this is more related to operational concerns, that is, the ongoing utility and operational costs of managing security. Once again, absent objectives, this will be difficult to measure.

The typical resources that must be managed are not significantly different than any other organizational department from a process perspective, although the resources will differ. They will comprise personnel, physical, and technical assets. Some of the decisions that the security manager may need to make could be based on the following issues:

- What would constitute effective and efficient resource management?
- What measures would be useful?
- Can existing resources be used to greater effect?
- What processes might be put in place that ensure effective and efficient use of resources?
- How would that process be monitored or measured?

One measure would be benchmarking against comparable organizations. If better security, that is, fewer losses related to security failures, is achieved at lower costs, the argument could be made that it is the result of better resource management. There could, of course, be other factors but the assertion could certainly be supported.

A useful measure could be resource utilization. For example, a measure of staff productivity could be used as an indicator of resource management. Reductions in the number of controls required to meet objectives would be a good indicator and an ongoing metric.

14.3.7 Performance Measurement—Monitoring and Reporting on Security Processes to Ensure that Organizational Objectives are Achieved

What decisions will the security manager need to make in regard to performance measurement? Assuming that it is apparent that it will be difficult to manage what is not being measured, decisions must be made about the state of performance measurement itself, based on information regarding some of the following questions:

- Are there adequate performance measures in place for strategic, tactical, and operational elements?
- Are all key controls monitored in some fashion that will indicate effectiveness?
- Are there measures of key control reliability?
- Will there be a clear and timely indication of control failure?
- Is the management process itself measured and monitored?

Answers to the foregoing will determine what actions must be taken.

14.4 INFORMATION SECURITY OPERATIONAL METRICS

Operational security personnel at various levels of the organization have different responsibilities and are required to make different decisions in the process of discharging their duties. The information needed for decision support will come from operations as opposed to strategic sources. As an example, consider a typical IT or information security manager who reports directly or indirectly to the CIO and often has a system administrator background. Activities are generally focused around operational issues involving the data center, network, and desk-top computers. Responsibilities usually include such things as firewall configuration and management, antimalware activities, patching, IDS/IPS-related activities, pen testing, vulnerability scanning, configuration management, compliance monitoring and enforcement, and incident response. There are many variations on this theme and perhaps many other specific tasks. Some will be more on the technical side, some less; some have greater responsibilities, and some less, but this encompasses the generality.

14.4.1 IT and Information Security Management

Many technical metrics are available in the IT environment. The larger problem is often too much data and too little information. What decisions does this manager make and what information is needed to ensure that an appropriate conclusion is arrived at? In terms of securing the IT environment, even with the vast amounts of data available, it is not an easy task to determine what might serve as useful metrics that meet the previously mentioned criteria of:

- Manageable
- Meaningful
- Actionable
- Unambiguous
- Reliable
- Timely
- Predictive

14.4.2 Compliance Metrics

Compliance with policy and technical standards is typically an area of responsibility for security managers. Procedural compliance is an important component of providing effective security. After all, even well-developed, tested procedural controls are only as good as the level of compliance.

The objective will normally be full and consistent compliance with mandated procedures. The decisions the manager must make related to compliance will generally revolve around either training and education or, alternatively, enforcement, and perhaps both. That is, it may require training in procedural controls to ensure that personnel know what to do or it may require enforcement activities to ensure that compliance takes place. It may also be necessary to measure the functionality of the procedure itself or determine that a particular function requires additional staff or resources.

For the manager to make decisions regarding which direction to take, specific additional information will be needed such as the actual level of compliance and whether individuals understand how to perform the control procedures. Without these metrics, it cannot be determined whether education or enforcement is required, or whether additional resources must be allocated. If the problem is indifference or carelessness, training might be a waste of time and money. In some cases, the best decision may be that a particular procedure is too cumbersome, impractical, or ineffective and must be changed.

To provide the information needed to make an informed decision in this case, we need to monitor or measure several elements including compliance levels, performance proficiency, and awareness levels. It may also be necessary to test and evaluate a procedure to measure functionality, efficiency, and appropriateness. There will be the varying degree of sensitivity and criticality of various activities, mandating different degrees of security effort to ensure that acceptable levels of risk are maintained. The assets involved will be subject to varying levels of risk exposure. Obviously, highly critical functions with significant exposures dependent on fully functioning controls will need better and more timely metrics or monitoring than those not so important. The reliability of the metrics themselves may also be an important consideration to factor into any decision process.

In this situation, the relevant information required for management decisions can include:

1. The criticality and sensitivity of assets involved
2. The level of risk the assets are exposed to
3. The state of compliance with the relevant procedure
4. The degree of procedural competence of personnel
5. The adequacy of resources
6. The reliability and accuracy of the metrics themselves
7. The functionality, efficiency, and appropriateness of the procedure

Let us consider how the required information may be acquired for each of these elements.

14.4.2.1 Criticality and Sensitivity

The level of compliance that may be satisfactory from a purely risk-based view will hinge on the level of criticality and/or sensitivity of the procedural activity under consideration. For procedures that are not critical or sensitive, a low level of procedural compliance may pose little risk. A broader issue may be the difficulty in structurally, operationally, and culturally mandating and enforcing vastly different levels of compliance. One risk is that personnel in the habit of cutting corners in performing some procedures are likely to do so in others as well.

For activities that are highly critical or sensitive, procedural compliance at anything less than one hundred percent will be unacceptable. Regardless, realities and limited resources may force priorities to be set for monitoring and metrics activities and, obviously, the highest concern will be for the most critical functions with significant degrees of exposure.

From a management standpoint, the level of criticality or sensitivity is required information for informed decision making. Although compliance failure for noncritical functions may still need to be dealt with and decisions made, the urgency will be less and the decisions required will not suffer from postponement.

Determining Criticality and Sensitivity. Various approaches can be used to determine the sensitivity and criticality of assets. Some form of asset classification is typically used and may also include performing business impact analysis or resource dependency assessment. Asset owners will need to be engaged to make these determinations.

Sensitivity can be measured by the extent to which there are adverse consequences, or impact, from unplanned disclosure. The typical approach is qualitative and quite subjective, often just ranked in three or four categories and having classifications such as confidential or in confidence, internal, and, perhaps, public or unclassified.

Criticality is the measure of how important a particular asset is to the business function and the consequences, or impact, of the loss of the use of the asset over time. This is usually measured in terms of some sort of ranking or by recovery time objectives (RTOs) based on a business impact assessment (BIA). It is obvious that for organizations that have not performed these classifications, it will be difficult to determine appropriate metrics for compliance since, they should arguably be related to the relative importance of the assets.

Sensitivity and criticality are not likely to change quickly and once determined by a classification process, change management should in most cases inform of any significant changes that would affect either of those dimensions. Nevertheless, either may change over time for reasons that will not be reflected in change management processes. For example, a database supporting a new service may initially only have a few items of sensitive information, but over time this may grow to a great deal of sensitive information, the compromise of which might pose a major risk. The growth in the quantity of information will not be subject to change management processes but the risk and potential impacts may increase substantially. Of course, this potential change should have been anticipated and compensatory activities undertaken at appropriate times, but it would be a prudent practice to plan some form of monitoring or review process to identify significant changes in risk for assets that may change in utility or value over time.

The decision that needs to be made will require information on:

1. The criticality or sensitivity of the resources. This can be in relative terms and approximated as low, medium, or high.
2. The control objectives in specific terms. What will the control accomplish and how will that be measured? For example, access control that will provide a 99.9% certainty that unauthorized access will not occur.
3. Will the controls considered provide the required level of certainty? This may require testing and/or statistical analysis to determine. For example, two layered, sequential controls with a 97% certainty of precluding unauthorized access and should provide the required 99.9% level of certainty.
4. The issue will then be, what combination of controls will provide this level of control at what cost? The costs will need to based on full life cycle TCO computations including acquisition, deployment, operation, maintenance, and testing.
5. Consideration must also be given to the fact that 99.9% certainty may mean that, on average, for each thousand events, an unauthorized individual may gain access, and the potential impact of that must be considered. Layered access controls will typically include authorizations as well, which in combination will raise this possibility to a more secure level.

14.4.2.2 Risk Exposure

The level of exposure will be a major factor in making decisions regarding compliance. If the level is very low for even a highly critical asset, the need for real-time metrics or monitoring is diminished. If, on the other hand, a critical asset is exposed to considerable risk, metrics or monitoring providing a high level of assurance of procedural compliance is warranted. In addition, it may also be justified to provide inherently strong controls such as dual control.

Risks. Risks are measured by risk assessments. These are, of course, also qualitative, subjective, and generally speculative. Nevertheless, relative risk to the assets under consideration will need to be known for effective decision support. Compliance with control procedures will be more critical for highly critical assets at substantial risk than for unimportant assets at little risk. Precision is not a necessity for decision making, just a process that will provide a consistent relative risk ranking in relation to other assets, since the objective is ensure the highest levels of compliance for the most critical assets at the most risk.

Risk will change over time or with the advent of new threats. A risk assessment combined with effective change management processes will often provide much of this information. But threats, vulnerabilities, and potential impacts will change and must be monitored.

14.4.2.3 The State of Compliance

The next element to consider is, what unit of measurement is useful for compliance? Percentage is a common measure and would seem obvious, but on consideration, it is inadequate. Ninety percent compliance with the steps required in a critical procedure can be fatal. For important or critical procedures, we need one hundred percent compliance with all the steps, one hundred percent of the time. Consequently, the metric does not need to be scalar, just binary; either the procedures are consistently followed or they are not.

It can be argued that a hundred percent compliance is unrealistic and high percentages might be adequate. For noncritical procedures, that might be sufficient, but in the case of heart surgery or flying jumbo jets, the argument is not persuasive and procedural failures have resulted in fatalities and huge negligence awards. Given potentially catastrophic consequences, few knowledgeable managers would accept high percentages of critical procedural compliance as acceptable.

If procedural compliance for critical systems is an absolute requirement, the unit of measurement can simply be yes or no, green or red. The issue then is how to acquire the information on an ongoing, real-time basis. Compliance with technical controls can generally be reliably automated but procedural controls pose a greater challenge. In many instances, procedures that interact with technical elements might provide the possibility of automated compliance metrics. In other instances, there may be no options other than with some form of monitoring such as CCTV. But direct observation or monitoring is inefficient and expensive and not the preferred option if other avenues exist. For trusted personnel, a common approach is the use of a signed manual checklist.

A variety of possibilities exist that can be considered for purely physical procedures. A typical physical access control procedure in highly sensitive or critical areas includes guards, ID badges, proximity cards, and sign-in logs. For handling critical backup tapes, procedures might include elements such as secure containers, identification and authentication of personnel transporting the tapes, and signed receipts, as well as CCTV recordings at pickup and delivery points. Although these sorts of procedures are common and generally effective, providing real-time metrics on compliance with the procedures is not straightforward and not typically done. The reality is that the compliance metrics in these cases involve after-the-fact reviews of logs and tapes, which is both troublesome and rarely performed unless there is a discovered incident.

Audits are the primary metric for compliance but suffer from the lack of timeliness. In the case of tapes, possible approaches could include secured GPS tracking devices or RFID chips and scanners at particular points. The required information could be gathered by a series of reporting checkpoints.

Compliance may pose the greatest challenge in terms of monitoring or metrics. Compliance monitoring or metrics, as previously mentioned, may be available from automated sources or may require physical or electronic observation in some fashion. Most organizations measure compliance primarily through audits, which may not provide information that is timely enough to meet risk management objectives. The decision on a reasonable level of monitoring will be based on the other foregoing elements. In cases with low proficiency of operating personnel, high risks, and high criticality or sensitivity, it is obvious that more monitoring and/or better metrics will be required. On the other hand, highly trained, experienced, personnel performing consistently and reliably over time may not require more than a periodic check to provide adequate levels of assurance.

Cultural, psychological, and organizational structural aspects will be a factor in the human dimension that in many circumstances can be quite important. Although these dimensions are likely to be too esoteric for most organizations, they must be considered when potential impacts can be catastrophic, such as dealing with nuclear weapons or the kinds of currency trading that bankrupted Barings Bank some years ago. Individual "world views" of personnel involved in highly critical or sensitive activities is an additional dimension that might be usefully considered. Employees with a "hierarchical" perspective are more likely to follow rules than individuals with an "individualist" outlook. Approaches to testing these dimensions are discussed in Appendix B.

Compliance measures for technical control procedures can generally be obtained from logs. The problem of reviewing the immense amounts of data from the typical logging activities can to some extent be addressed through so-called SIM log reading and correlation tools. In the process of essentially data mining the logs, these tools can be configured in a number of ways that can address many metrics requirements.

Measures of compliance with physical procedures may require, in addition to periodic audits, random inspections, security reviews, video monitoring, supervisory oversight, or guards.

Compliance metrics can often be obtained from procedures that interface with technology. An example would be the procedures for hardening a server. Activities can be logged and the required steps can be verified or the configuration can be tested automatically to conform to specifications. Checking the configuration does not ensure that the steps were followed exactly but may nevertheless be adequate to ensure the desired outcomes.

We can monitor a process by some means to ensure that all steps required by a procedure are followed. We can provide some type of outcome metrics to ensure that we achieve the desired results. For processes that engender risks if steps are not followed (such as mixing nitroglycerine to make dynamite or flying jets), either monitoring at critical stages or creating metrics that warn of impending danger can be used.

Are there elements of managing security that create hazardous situations at various points, even if final outcomes can be verified as being satisfactory? A number of situations come to mind and the answer is in the affirmative. An example would be the event of a breach not being responded to according to the correct procedures and, consequently, escalating into a full-blown catastrophe. Events at a major financial institution involving the Slammer worm serves to illustrate this possibility.

Case Study. Personnel monitoring the network operations center (NOC) noticed unusual network activity on a Sunday evening. Deciding there was no imminent danger and certain that they could handle any eventuality, they decided to watch the event. By the early morning, traffic continued to increase at both the main facility and then suddenly began to grow dramatically at the mirror site hundreds of miles away. By 7 a.m., they notified senior IT managers that there was a problem and the network was becoming saturated. An hour later, when the external CIRT team arrived, the network was totally inoperative and it was determined that they had in fact been compromised by the Slammer worm. The CIRT team manager informed the network manager that Slammer was memory resident and restarting the entire network and mirror facility would resolve the issue. The manager stated that he did not have the authority to do so and it would require the CIO to issue that instruction. The CIO could not be located and current phone numbers were kept in an emergency paging system that required network access. When asked about the disaster recovery plan and what it had to say regarding declaration criteria, three different plans were produced that had been prepared by teams in different parts of the organization, unbeknownst to each other. The final resolution ultimately required the CEO, who was also not immediately available, to finally issue instructions the next morning to shut the nonfunctioning network down. Over thirty thousand people could not perform their work and the institution was inoperative for a full day and a half. The final direct costs were estimated by the postmortem team to exceed fifty million dollars despite the stonewalling and lack of cooperation from most employees fearful of being found somehow at fault.

The author managed the postmortem team that found literally hundreds of deficient processes, a dysfunctional culture, and an array of useless metrics, in addition to a fatally flawed lack of adequate governance. For those monitoring the NOC, metrics indicated a problem but they were not sufficiently meaningful to the employees for them to make any active decisions, much less the correct ones. Either better metrics or greater proficiency of the personnel could have resolved the issue quickly in the initial stages of the incident before it became a problem. Better governance would have vested adequate authority in the network manager to take the appropriate action. Better governance would have insisted that either the vulnerability patches issued a full two months prior were applied or suitable compensatory controls were implemented to address a well-known threat. Even marginally effective risk management would have insisted that a flat network with no segmentation was unacceptable and that DR/BCP was an integrated and tested activity.

The conclusions that can be reached that are relevant to metrics are that *data is not information* and that incomprehensible information is just data and useless. It also illustrates that no matter how good the metrics and monitoring providing decision support, it is useless to someone not empowered to make decisions.

As a consequence, to develop useful metrics, it must be clear what decisions must be made, by whom, and what information is needed to make them. It is apparent from this analysis that management metrics will typically require a variety of information from different sources that must then be synthesized to provide meaningful information needed for making decisions about what actions are required.

14.4.2.4 Personnel Competence

Metrics around the competence of personnel to perform complex procedures reliably and consistently will be important as well. The metrics indicating high levels of personnel proficiency, commitment, integrity, and reliability can to some indeterminate extent, under some conditions, reduce the need for real-time continuous monitoring or metrics. Complex procedures may require training and/or periodic refresher courses if metrics indicate inadequate proficiency. The decisions a manager might need to make for the amount and type of training will benefit from metrics in terms of personnel awareness, proficiency, and reliability.

Proficiency and Awareness. Proficiency in performing procedures can be tested in a variety of fairly obvious ways. Written tests, quizzes, direct observation, or tracking errors or omissions are some of the possibilities. Performance records over a period of time may be useful measures as well. Security awareness can be measured by simple questionnaires or quizzes administered periodically.

14.4.2.5 Resource Adequacy

The adequacy of resources can be determined by observation, by analysis, or by questioning the individuals involved. Inadequacy of resources is often a function of misallocation, underutilization, or a productivity issue. Standard approaches can be utilized to determine which is the cause.

In the case of personnel, a typical approach to measurement would be detailed time sheets to determine how time is utilized. It is likely that high-value individuals are spending too much time on low-value activities, which would support the decision for better task and time management or, possibly, outsourcing low-value activities.

14.4.2.6 Metrics Reliability

Metrics on the reliability and accuracy of the metrics themselves will be needed in order to make appropriate decisions. If the metrics of a critical function are only 90% reliable and 60% accurate, additional actions will be required to achieve high assurance levels since the information can only be relied on 45% of the time. Another consideration for the decision maker will be whether the metrics will indicate false positives or false negatives, or whether in some cases they will fail to show anything.

Metrics need to be tested periodically to provide assurance of functionality, reliability, and accuracy. Depending on the metric, reliability can be measured in a number of ways. Repeatability and consistency are useful for measuring both reliability and accuracy. For technical metrics, testing can usually be automated to provide assurance of consistency. Other gauges of reliability can range from statistical sampling and analysis over a large number of measurements to comparison with some sort of standard.

Measures may be evaluated against outcomes to determine reliability and consistency. That is to say, if a particular metric value consistently results in the same outcome, it is reliable. If every time the fuel gauge shows empty, the vehicle ceases operating, there is a level of assurance of the reliability of the gauge. Conversely, if periodically it shows an amount of fuel when there is none, prudence dictates that some other metric should be used such as a dipstick in the gas tank.

14.4.2.7 Procedure Functionality, Efficiency, and Appropriateness

Finally, it is necessary to consider the possibility that if there are problems with compliance then a particular procedural control is poorly designed, difficult to perform, or inefficient. An analysis of the control objectives may be needed to ensure that they are properly aligned with business and security requirements. Research indicates that, more often than not, poor performance by employees is the result of bad process design. As a result, the procedure may need to be tested and evaluated from a process standpoint and, perhaps, be redesigned or possibly automated.

In most organizations, procedures that more or less work are typically not subject to review. If compliance is found to be a problem, a good candidate for the root cause will be poor procedures. If control objectives have not been defined and the assets not classified, it will be difficult to determine appropriate procedural controls, however.

With the assumption that procedural controls are documented, one measure is the completeness, accuracy, consistency, and conformance to standards (and, therefore, policy) of the procedure itself. Another measure is the ability of someone unfamiliar with the procedure to accurately accomplish the task using the written procedure.

14.4.2.8 Tactical Performance Measures

Historical data, to the extent it is available, can be useful in determining how effective existing performance measures are. Analysis of prior events and postmortems can yield information on issues such as:

- How long it has taken to detect security incidents
- The overall impacts and remediation costs
- Incident recovery times

14.4.2.9 Key Control Effectiveness

Key control effectiveness must be periodically tested as it tends to degrade over time. In other words, do the controls continue to meet the control objectives? Procedural controls are particularly susceptible to being circumvented or degraded, resulting in decreased effectiveness. If these are key controls protecting critical assets, they may need additional controls or to be strengthened commensurate with criticality or sensitivity.

14.4.2.10 Control Reliability

A critical aspect of controls is reliability. Depending on asset or process criticality, unreliable controls pose a serious risk, particularly if they are not monitored in real time. Automated controls are typically more reliable than procedural ones but this should not be assumed. All key controls must be tested periodically and should have monitoring processes to warn of failure in real time. The processes used for monitoring will require testing as well to ensure that they signal control failure reliably.

14.4.2.11 Control Failure

Methods and processes must exist to signal key control failure on a real-time basis. The monitoring and notification processes must be tested on a regular basis to ensure that they operate as intended. Key control failure is information essential for effective security management.

14.4.2.12 Management Effectiveness

The effectiveness of information security management will ultimately be measured by whether or not defined objectives are achieved. As with any other process, increased productivity and greater resource utilization can be indicators of management effectiveness. Another marker might be trends in strategic, proactive security activities as opposed to purely tactical, reactive ones.

Chapter 15

Incident Management and Response Metrics

Incident management and response are the last steps in risk management and the final barrier to what may become an unmitigated disaster. The decisions that must be made require accurate and timely information and will probably include many of the following:

- Is it actually an incident?
- What kind of incident is it?
- Is it a security incident?
- What is the severity level?
- Are there multiple events and/or impacts?
- Will they need triage?
- What is the most effective response?
- What immediate actions must be taken?
- Which incident response teams and other personnel must be mobilized?
- Who must be notified?
- Who is in charge?
- Is it becoming a disaster?

There may be assistance available for making some of the necessary decisions, such as event detection and correlation tools. Specific suggestions for metrics are difficult as individual circumstances are highly variable. Nevertheless, this is a vital area for most organizations and we will attempt to identify the different types of information needed and some possibilities for acquiring it. Once again, management information requirements will differ from operational or incident response needs.

15.1 INCIDENT MANAGEMENT DECISION SUPPORT METRICS

The answers to the questions above will determine how to proceed in the event of an incident and may prove critical to the organization. Let us consider them individually and see what information is required as well as possible sources and metrics.

15.1.1 Is it Actually an Incident?

Nontechnical personnel encountering unusual events on the network or on their computer may suspect that an incident is taking place and report it, perhaps to the help desk. Help desk staff unfamiliar with the situation may in turn escalate it to IT or security, or ignore it. If the situation is unfamiliar to security staff or IT, how will they determine if the situation constitutes an incident and is in fact posing a risk? What must be done to clarify whether just an individual desktop machine has a problem or it is a manifestation of a bigger event? It is likely to prove useful to have a defined, systematic approach of troubleshooting to quickly determine these answers so that proper response can be initiated. As in medical situations, early detection and proper diagnosis are key to preventing serious impacts. In the majority of organizations, general awareness education about incidents and specific training for key personnel is the proven method of improving incident detection, management, and response.

Depending on what is monitored and the types of metrics available, there will be information, which, if accessible, will be helpful in incident diagnosis. For example, incidents involving intrusions into a network may be detected by network intrusion detection systems (NIDS) or host IDS (HIDS) or other tools such as security information management (SIM) tools checking logs. There may be possibilities for detecting anomalies when the ranges of normal operations are known, in a manner similar to anomaly-based IDSs. In other words, unusual levels of traffic or certain operations at unusual times can be an indication that something is amiss. This is true of physical access as well, and having visibility into existing physical access controls is essential and, surprisingly, in most organizations, unusual.

A simple and effective monitoring tool is to simply correlate physical presence with log-in location. Obviously, if someone not shown to be on the premises by physical access controls is found to be logging into the system from the premises (i.e., not remotely) either there is a failure of the access controls or there is some other security incident occurring.

The information needed to determine whether an event constitutes an incident will require personnel monitoring resources having suitable skills as well as effective communication channels. If skills are not adequate, the decision will be to either provide training or acquire personnel with the necessary skills. If who needs to communicate what to whom, when, and how is not understood or the channels are not available, the decisions will revolve around addressing those issues.

Understanding that this information is required to make appropriate decisions makes it straightforward to determine what metrics or monitoring are needed for management.

The skills needed must be determined and then skills assessment testing is required to determine the gap that must be addressed. Communication channels can be tested by simulated incidents, perhaps starting with a table-top walk-through to ensure that personnel have the right information and understand their responsibilities. The metric will be binary, that is, personnel either have the skills and know what to do or they do not.

15.1.2 What Kind of Incident Is It?

There are generally few metrics likely to directly identify the type of incident except monitoring and diagnostic capabilities. An example is a physical event that will have technical manifestations such as lost connectivity when a ditch digging machine severs the fiber link to the data center or a wiring closet burns up. Whether the organization considers these to be security events varies, although, arguably, this will impact availability, which usually has security implications of one sort or another. If emergency services cannot be contacted, air traffic control cannot communicate with aircraft, or credit cards cannot be authorized, it is difficult to argue that availability is not a security issue.

Validating that an incident has occurred, as in the foregoing paragraph, will generally result in the determination of the kind of incident it is. Additional information will usually be required such as the scope and possible impact of the incident as well.

15.1.3 Is It a Security Incident?

There is little consensus on exactly what constitutes a "security" event, although many organizations have developed internal criteria and definitions. Organizations often consider the cause of an incident to be determinative; that is, a deliberate disruptive act would be a security matter, whereas an accident would not. This distinction suffers from the fact that it is easy to imagine many accidental situations that can have major security implications and impacts. It may be more prudent to consider a security incident as *any event that has the potential of compromising security or elevating risk, regardless of cause.*

15.1.4 What Is the Severity Level?

Severity of an incident must be determined quickly and, hopefully, with a degree of accuracy. Declaring a full-fledged disaster as the result of a minor incident is not likely to be a good career move and neither is failing to declare one and reacting appropriately when there actually is one.

Severity levels must be defined and agreed upon, and personnel must be educated or trained to make the determination. Authority to declare the various severity levels must be assigned and escalation procedures defined. Other preconditions exist such as information asset classification, which is required so that the criticality and/or sensitivity of affected assets can be quickly determined, which, along with the level of impact, will be largely determinative of severity level.

Once again, good diagnostics will be the key to efficiently gathering the needed information to arrive at a conclusion and initiate appropriate action.

15.1.5 Are there Multiple Events and/or Impacts?

Incidents can aggregate and/or cascade. That is, one threat can affect multiple resources concurrently, or an incident can initiate a chain of events, causing a "cascade" of failures, the so-called domino effect. It will be critical to determine the scope of the impact or whether there are other resources at risk as a result of an event. Metrics and monitoring are often helpful in assessing scope but intimate knowledge of systems, networks, personnel, or facilities are likely to be needed to assess likely "knock-on" eventualities.

15.1.6 Will an Incident Need Triage?

Multiple events that exceed the organization's incident response capacity to address them all will require triage to determine which issues to deal with; which to ignore, either because they are not serious or there is no ability to address them effectively; and in which order. This capability requires substantial expertise; systems, personnel, and, possibly, facilities knowledge; a variety of real-time operational metrics about what is working and what is not; performance impacts; and so on. For purely technical events, the typical range of data being monitored in the NOC may be adequate provided there is the expertise to interpret it correctly.

15.1.7 What is the Most Effective Response?

Determining the most effective response to a security incident requires the right information and knowledge of the available options. For example, if an attacker has breached perimeter security and raised an intrusion detection alert, what action should be taken? Perhaps the network is segmented and the attack can be isolated. Perhaps the intruder can be blocked at the firewall or, possibly, more drastic action is required such as terminating the connection to the Internet. Without adequate information and an understanding of the systems and architectures, it will be difficult to determine the least disruptive response consistent with security.

Physical compromise such as theft of proprietary information or indications of embezzlement by an insider will present even more challenging response issues. Often, the main metrics and sources of information for these types of events will be technical or accounting forensics.

15.1.8 What Immediate Actions Must be Taken?

Some incidents will require immediate action to avoid serious consequences. HIDS or NIDS inside the perimeter signaling an intrusion certainly qualifies. Besides validating that it is in fact an intrusion, operational metrics indicating the scope and nature of activity are critical to deciding the nature and extent of action required.

In many organizations in which operations and traffic follows consistent patterns, anomalies may be a useful metrics to warn of incipient incidents.

15.1.9 Which Incident Response Teams and Other Personnel Must be Mobilized?

The type and nature of an incident must be determined to make decisions about which teams or personnel will be required to deal with it.

15.1.10 Who Must be Notified?

Utilizing defined severity criteria will provide the input into the declaration criteria, which,if properly developed, defines who has what authority and who must be notified.

15.1.11 Who Is in Charge?

Indecision and lack of clear authority to take necessary actions can and has transformed and incident into a disaster.

15.1.12 Is it Becoming a Disaster?

It is clear that one of the metrics the information security manager must develop from an incident management and response perspective is the level of skill and expertise available at a given point in time to address events.

Chapter **16**

Conclusion

For the stalwart reader that has come this far, it should be evident that developing and implementing information security governance in situations where little exists or where a significant overhaul is required is no minor endeavor. Nevertheless, moving into the future, there are not many good options and organizations will have few prudent choices. The combination of regulatory, legal, and credit industry pressures coupled with ever more spectacular security compromises will make it increasingly difficult to continue the ad-hoc, reactive, point solution approaches that are still the norm. It has become evident that the historical approach of seeking the substantial benefits offered by information technology and global interconnectivity cannot continue to operate "on the cheap" and must be properly governed and managed. As with any major organizational activity, this requires commitment and resources. More importantly, it also requires a systematic, organized, engineered approach managed by competent professionals, coupled with good business sense and the "soft" interpersonal skills that are the mark of effective managers.

It is essential to understand that effective governance is not possible without metrics, and the converse is true as well. The maxim that "what gets measured gets done" is well founded. It is clear that the way forward for security is not significantly different from that of any other nascent discipline. The processes that give rise to reasonable levels of assurance require integration into the fundamental objectives of the enterprise at the highest levels. It must be the concomitant with setting organizational goals and it must become the nonnegotiable requirement of the exercise of due care. Until such time, information security will be characterized by the haphazard, reactive point solutions typical of security departments generally operating behind the power curve and all too often in crisis mode.

Appendix A

SABSA Business Attributes and Metrics

Business attribute	Attribute explanation	Metric type	Suggested measurement approach
User attributes. These attributes are related to the user's experience of interacting with the business system.			
Accessible	Information to which the user is entitled to gain access should be easily found and accessed by that user.	Soft	Search tree depth necessary to find the information
Accurate	The information provided to users should be accurate within a range that has been preagreed upon as being applicable to the service being delivered.	Hard	Acceptance testing on key data to demonstrate compliance with design rules
Anonymous	For certain specialized types of service, the anonymity of the user should be protected.	Hard Soft	Rigorous proof of system functionality Red team review*
Consistent	The way in which log-in, navigation, and target services are presented to the user should be consistent across different times, locations, and channels of access.	Hard Soft	Conformance with design style guides Red team review

Business attribute	Attribute explanation	Metric type	Suggested measurement approach
Current	Information provided to users should be current and kept up to date, within a range that has been preagreed upon as being applicable for the service being delivered.	Hard	Refresh rates at the data source and replication of refreshed data to the destination
Duty-segregated	For certain sensitive tasks, the duties should be segregated so that no user has access to both aspects of the task.	Hard	Functional testing
Educated and aware	The user community should be educated and trained so that they can embrace the security culture There should be sufficient user awareness of security issues so that behavior of users is compliant with security policies.	Soft	Competence surveys
Informed	The user should be kept fully informed about services, operating procedures, operational schedules, planned outages, and so on.	Soft	Focus groups or satisfaction surveys
Motivated	The interaction with the system should add positive motivation to the user to complete the business tasks at hand.	Soft	Focus groups or satisfaction surveys
Protected	The user's information and access privileges should be protected against abuse by other users or by intruders.	Soft	Penetration test. (Could be regarded as "hard," but only if a penetration is achieved. Failure to penetrate does not mean that penetration is impossible.)
Reliable	The services provided to the user should be delivered at a reliable level of quality.	Soft	A definition of "quality" is needed against which to compare.
Responsive	The users obtain a response within a satisfactory period of time that meets their expectations.	Hard	Response time

Business attribute	Attribute explanation	Metric type	Suggested measurement approach
Supported	When a user has problems or difficulties in using the system or its services, there should be a means by which the user can receive advice and support so that the problems can be resolved to the satisfaction of the user.	Soft	Focus groups or satisfaction surveys. Independent audit and review against Security Architecture Capability Maturity Model†
Timely	Information is delivered or made accessible to the user at the appropriate time or within the appropriate time period.	Hard	Refresh rates at the data source and replication of refreshed data to the destination
Transparent	Providing full visibility to the user of the logical process but hiding the physical structure of the system (as a url hides the actual physical locations of Web servers).	Soft	Focus groups or satisfaction surveys. Independent audit and review against Security Architecture Capability Maturity Model†
Usable	The system should provide easy-to-use interfaces that can be navigated intuitively by a user of average intelligence and training level (for the given system). The user's experience of these interactions should be at best interesting and at worst neutral.	Soft	Numbers of "clicks" or keystrokes required. Conformance with industry standards, e.g., color palettes. Feedback from focus groups.
Management attributes. This group of attributes is related to the ease and effectiveness with which the business system and its services can be managed.			
Automated	Wherever possible (and depending upon cost/benefit factors) the management and operation of the system should be automated.	Soft	Independent design review

Business attribute	Attribute explanation	Metric type	Suggested measurement approach
Change-managed	Changes to the system should be properly managed so that the impact of every change is evaluated and the changes are approved in advance of being implemented.	Soft	Documented change management system, with change management history, evaluated by independent audit
Controlled	The system should at all times remain in the control of its managers. This means that the management will observe the operation and behavior of the system, will make decisions about how to control it based on these observations, and will implement actions to exert that control.	Soft	Independent audit and review against Security Architecture Capability Maturity Model†
Cost-effective	The design, acquisition, implementation, and operation of the system should be achieved at a cost that the business finds acceptable when judged against the benefits derived.	Hard	Individual budgets for the phases of development and for ongoing operation, maintenance and support
Efficient	The system should deliver the target services with optimum efficiency, avoiding wastage of resources.	Hard	A target efficiency ratio based on (Input value)/(Output value)
Maintainable	The system should capable of being maintained in a state of good repair and effective, efficient operation. The actions required to achieve this should feasible within the normal operational conditions of the system.	Soft	Documented execution of a preventive maintenance schedule for both hardware and software, correlated against targets for continuity of service, such as mean time between failures (MTBF)
Measured	The performance of the system should be measured against a variety of desirable performance targets so as to provide feedback information to support the management and control process.	Hard	Documented tracking and reporting of a portfolio of conventional system performance parameters, together with other attributes from this list

Business attribute	Attribute explanation	Metric type	Suggested measurement approach
Supportable	The system should be capable of being supported in terms of both the users and the operations staff, so that all types of problems and operational difficulties can be resolved.	Hard	Fault-tracking system providing measurements of MTBF, MTTR (mean time to repair), and maximum time to repair, with targets for each parameter
Operational attributes. These attributes describe the ease and effectiveness with which the business system and its services can be operated.			
Available	The information and services provided by the system should be available according to the requirements specified in the service-level agreement (SLA).	Hard	As specified in the SLA
Continuous	The system should offer "continuous service." The exact definition of this phrase will always be subject to a SLA.	Hard	Percentage up-time correlated versus scheduled and/or unscheduled downtime, or MTBF, or MTTR
Detectable	Important events must be detected and reported.	Hard	Functional testing
Error-free	The system should operate without producing errors.	Hard	Percentage or absolute error rates (per transaction, per batch, per time period, etc.)
Interoperable	The system should interoperate with other similar systems, both immediately and in the future, as intersystem communication becomes increasingly a requirement.	Hard	Specific interoperability requirements
Monitored	The operational performance of the system should be continuously monitored to ensure that other attribute specifications are being met. Any deviations from acceptable limits should be notified to the systems management function.	Soft	Independent audit and review against Security Architecture Capability Maturity Model†

Business attribute	Attribute explanation	Metric type	Suggested measurement approach
Productive	The system and its services should operate so as to sustain and enhance productivity of the users, with regard to the business processes in which they are engaged.	Hard	User output targets related to specific business activities
Recoverable	The system should be able to be recovered to full operational status after a breakdown or disaster, in accordance with the SLA.	Hard	As specified in the SLA.
Risk management attributes. These attributes describe the business requirements for mitigating operational risk. This group most closely relates to the "security requirements" for protecting the business.			
Access-controlled	Access to information and functions within the system should be controlled in accordance with the authorized privileges of the party requesting the access. Unauthorized access should be prevented.	Hard	Reporting of all unauthorised access attempts, including number of incidents per period, severity, and result (did the access attempt succeed?)
Accountable	All parties having authorized access to the system should be held accountable for their actions.	Soft	Independent audit and review against Security Architecture Capability Maturity Model† with respect to the ability to hold accountable all authorized parties
Assurable	There should be a means to provide assurance that the system is operating as expected and that all of the various controls are correctly implemented and operated.	Hard	Documented standards exist against which to audit
		Soft	Independent audit and review against Security Architecture Capability Maturity Model†

Business attribute	Attribute explanation	Metric type	Suggested measurement approach
Assuring honesty	Protecting employees against false accusations of dishonesty or malpractice.	Soft	Independent audit and review against Security Architecture Capability Maturity Model[†] with respect to the ability to prevent false accusations that are difficult to repudiate
Auditable	The actions of all parties having authorized access to the system, and the complete chain of events and outcomes resulting from these actions, should be recorded so that this history can be reviewed. The audit records should provide an appropriate level of detail, in accordance with business needs.	Soft	Independent audit and review against Security Architecture Capability Maturity Model[†]
	The actual configuration of the system should also be capable of being audited so as to compare it with a target configuration that represents the implementation of the security policy that governs the system.	Hard	Documented target configuration exists under change control with a capability to check current configuration against this target
		Soft	Independent audit and review against Security Architecture Capability Maturity Model[†]
Authenticated	Every party claiming a unique identity (i.e., a claimant) should be subject to a procedure that verifies that the party is indeed the authentic owner of the claimed identity.	Soft	Independent audit and review against Security Architecture Capability Maturity Model[†] with respect to the ability to authenticate successfully every claim of identity
Authorized	The system should allow only those actions that have been explicitly authorized.	Hard	Reporting of all unauthorized actions, including number of incidents per period, severity, and result (did the action succeed?)

Business attribute	Attribute explanation	Metric type	Suggested measurement approach
Authorized (*cont.*)		Soft	Independent audit and review against Security Architecture Capability Maturity Model† with respect to the ability to detect unauthorized actions
Capturing new risks	New risks emerge over time. The system management and operational environment should provide a means to identify and assess new risks (new threats, new impacts, or new vulnerabilities).	Hard	Percentage of vendor-published patches and upgrades actually installed
		Soft	Independent audit and review against Security Architecture Capability Maturity Model† of a documented risk assessment process and a risk assessment history
Confidential	The confidentiality of (corporate) information should be protected in accordance with security policy. Unauthorized disclosure should be prevented.	Hard	Reporting of all disclosure incidents, including number of incidents per period, severity, and type of disclosure
Crime-free	Cyber-crime of all types should be prevented.	Hard	Reporting of all incidents of crime, including number of incidents per period, severity, and type of crime
Flexibly secure	Security can be provided at various levels, according to business need. The system should provide the means to secure information according to these needs, and may need to offer different levels of security for different types of information (according to security classification).	Soft	Independent audit and review against Security Architecture Capability Maturity Model†

Business attribute	Attribute explanation	Metric type	Suggested measurement approach
Identified	Each entity that will be granted access to system resources and each object that is itself a system resource should be uniquely identified (named) such that there can never be confusion as to which entity or object is being referenced.	Hard	Proof of uniqueness of naming schemes
Independently secure	The security of the system should not rely upon the security of any other system that is not within the direct span of control of this system.	Soft	Independent audit and review against Security Architecture Capability Maturity Model† of technical security architecture at conceptual, logical, and physical layers
In our sole possession	Information that has value to the business should be in the possession of the business, stored and protected by the system against loss (as in no longer being available) or theft (as in being disclosed to an unauthorised party). This will include information that is regarded as "intellectual property."	Soft	Independent audit and review against Security Architecture Capability Maturity Model†
Integrity-assured	The integrity of information should be protected to provide assurance that it has not suffered unauthorized modification, duplication, or deletion.	Hard	Reporting of all incidents of compromise, including number of incidents per period, severity, and type of compromise
		Soft	Independent audit and review against Security Architecture Capability Maturity Model† with respect to the ability to detect integrity compromise incidents

Business attribute	Attribute explanation	Metric type	Suggested measurement approach
Non-repudiable	When one party uses the system to send a message to another party, it should *not* be possible for the first party to falsely deny having sent the message, or to falsely deny its contents.	Hard	Reporting of all incidents of unresolved repudiations, including number of incidents per period, severity, and type of repudiation
		Soft	Independent audit and review against Security Architecture Capability Maturity Model† with respect to the ability to prevent repudiations that cannot be easily resolved
Owned	There should be an entity designated as "owner" of every system. This owner is the policy maker for all aspects of risk management with respect to the system, and exerts the ultimate authority for controlling the system.	Soft	Independent audit and review against Security Architecture Capability Maturity Model† of the ownership arrangements and of the management processes by which owners should fulfil their responsibilities, and of their diligence in so doing
Private	The privacy of (personal) information should be protected in accordance with relevant privacy or "data protection" legislation, so as to meet the reasonable expectation of citizens for privacy. Unauthorized disclosure should be prevented.	Hard	Reporting of all disclosure incidents, including number of incidents per period, severity, and type of disclosure
Trustworthy	The system should be able to be trusted to behave in the ways specified in its functional specification and should protect against a wide range of potential abuses.	Soft	Focus groups or satisfaction surveys researching the question "Do you trust the service?"

Business attribute	Attribute explanation	Metric type	Suggested measurement approach
Legal and regulatory attributes. This group of attributes describes the business requirements for mitigating operational risks that have a specific legal or regulatory connection.			
Admissible	The system should provide forensic records (audit trails and so on) that will be deemed to be "admissible" in a court of law, should that evidence ever need to be presented in support of a criminal prosecution or a civil litigation.	Soft	Independent audit and review against Security Architecture Capability Maturity Model† by computer forensics expert
Compliant	The system should comply with all applicable regulations, laws, contracts, policies, and mandatory standards, both internal and external.	Soft	Independent compliance audit with respect to the inventories of regulations, laws, policies, etc.
Enforceable	The system should be designed, implemented and operated such that all applicable contracts, policies, regulations, and laws can be enforced by the system.	Soft	Independent review of: (1) inventory of contracts, policies, regulations and laws for completeness, and (2) enforceability of contracts, policies, laws, and regulations on the inventory
Insurable	The system should be risk-managed to enable an insurer to offer reasonable commercial terms for insurance against a standard range of insurable risks	Hard	Verify against insurance quotations
Legal	The system should be designed, implemented, and operated in accordance with the requirements of any applicable legislation. Examples include data protection laws, laws controlling the use of cryptographic technology, laws controlling insider dealing on the stock market, and laws governing information that is considered racist, seditious, or pornographic.	Soft	Independent audit and review against Security Architecture Capability Maturity Model.† Verification of the inventory of applicable laws to check for completeness and suitability

Business attribute	Attribute explanation	Metric type	Suggested measurement approach
Liability-managed	The system services should be designed, implemented and operated so as to manage the liability of the organization with regard to errors, fraud, malfunction, and so on. In particular, the responsibilities and liabilities of each party should be clearly defined.	Soft	Independent legal expert review of all applicable contracts, SLAs, etc.
Regulated	The system should be designed, implemented, and operated in accordance with the requirements of any applicable regulations. These may be general (such as safety regulations) or industry-specific (such as banking regulations).	Soft	Independent audit and review against Security Architecture Capability Maturity Model†. Verification of the inventory of applicable regulations to check for completeness and suitability
Resolvable	The system should be designed, implemented and operated in such a way that disputes can be resolved with reasonable ease and without undue impact on time, cost, or other valuable resources.	Soft	Independent audit and review against Security Architecture Capability Maturity Model† by legal expert
Time-bound	Meeting requirements for maximum or minimum periods of time, for example, a minimum period for records retention or a maximum period within which something must be completed.	Hard	Independent functional design review against specified functional requirements
Technical strategy attributes. This group of attributes describes the needs for fitting into an overall technology strategy.			
Architecturally open	The system architecture should, wherever possible, not be locked into specific vendor interface standards and should allow flexibility in the choice of vendors and products, both initially and in the future.	Soft	Independent audit and review against Security Architecture Capability Maturity Model† of technical architecture (conceptual, logical, and physical)

Business attribute	Attribute explanation	Metric type	Suggested measurement approach
COTS/GOTS compliant	Wherever possible, the system should utilize commercial off-the-shelf or government off-the-shelf components, as appropriate.	Soft	Independent audit and review against Security Architecture Capability Maturity Model† of technical architecture (conceptual, logical, and physical)
Extendable	The system should be capable of being extended to incorporate new functional modules as required by the business.	Soft	Independent audit and review against Security Architecture Capability Maturity Model† of technical architecture (conceptual, logical & physical)
Flexible & Adaptable	The system should be flexible and adaptable to meet new business requirements as they emerge.	Soft	Independent audit and review against Security Architecture Capability Maturity Model† of technical architecture (conceptual, logical, and physical)
Future-proof	The system architecture should be designed as much as possible to accommodate future changes in both business requirements and technical solutions.	Soft	Independent audit and review against Security Architecture Capability Maturity Model† of technical architecture (conceptual, logical, and physical)
Legacy-sensitive	A new system should be able to work with any legacy systems or databases with which it needs to interoperate or integrate.	Soft	Independent audit and review against Security Architecture Capability Maturity Model† of technical architecture (conceptual, logical, and physical)
Migrateable	There should be a feasible, manageable migration path, acceptable to the business users, that moves from an old system to a new one, or from one released version to the next.	Soft	Independent audit and review against Security Architecture Capability Maturity Model† of technical architecture (conceptual, logical, and physical)

Business attribute	Attribute explanation	Metric type	Suggested measurement approach
Multisourced	Critical system components should be obtainable from more than one source, to protect against the risk of the single source of supply and support being withdrawn.	Soft	Independent audit and review against Security Architecture Capability Maturity Model† of technical architecture at the component level
Scaleable	The system should be scaleable to the size of user community, data storage requirements, processing throughput, and so on that might emerge over the lifetime of the system.	Soft	Independent audit and review against Security Architecture Capability Maturity Model† of technical architecture (conceptual, logical, and physical)
Simple	The system should be as simple as possible, since complexity only adds further risk.	Soft	Independent audit and review against Security Architecture Capability Maturity Model† of technical architecture (conceptual, logical, and physical)
Standards compliant	The system should be designed, implemented and operated to comply with appropriate technical and operational standards.	Soft	Independent audit and review of: (1) the inventory of standards to check for completeness and appropriateness, and (2) compliance with standards on the inventory
Traceable	The development and implementation of system components should be documented so as to provide complete two-way traceability. That is, every implemented component should be justifiable by tracing back to the business requirements that led to its inclusion in the system, and it should be possible to review every business requirement and demonstrate which of the implemented system components are there to meet this requirement.	Soft	Independent expert review of documented traceability matrices and trees

Business attribute	Attribute explanation	Metric type	Suggested measurement approach
Upgradeable	The system should be capable of being upgraded with ease to incorporate new releases of hardware and software.	Soft	Independent audit and review against Security Architecture Capability Maturity Model[†] of technical architecture (conceptual, logical, and physical)
Business strategy attributes. This group of attributes describes the needs for fitting into an overall business strategy.			
Brand enhancing	The system should help to establish, build, and support the brand of the products or services based upon this system.	Soft	Market surveys
Business-enabled	Enabling the business and fulfilling business objectives should be the primary driver for the system design.	Soft	Business management focus group
Competent	The system should protect the reputation of the organization as being competent in its industry sector	Soft	Independent audit, or focus groups, or satisfaction surveys
Confident	The system should behave in such a way as to safeguard confidence placed in the organization by customers, suppliers, shareholders, regulators, financiers, the marketplace, and the general public.	Soft	Independent audit, or focus groups, or satisfaction surveys
Credible	The system should behave in such a way as to safeguard the credibility of the organization.	Soft	Independent audit, or focus groups, or satisfaction surveys
Culture-sensitive	The system should be designed, built, and operated with due care and attention to cultural issues relating to those who will experience the system in any way. These issues include such	Soft	Independent audit and review of (1) the inventory of requirements in this area to check for completeness and

Business attribute	Attribute explanation	Metric type	Suggested measurement approach
Culture-sensitive (*cont.*)	matters as religion, gender, race, nationality, language, dress code, social customs, ethics, politics, and the environment. The objective should be to avoid or minimize offence or distress caused to others.		appropriateness, and (2) compliance of system functionality with this set of requirements
Enabling time-to-market	The system architecture and design should allow new business initiatives to be delivered to the market with minimum delay.	Soft	Business management focus group
Governable	The system should enable the owners and executive managers of the organization to control the business and to discharge their responsibilities for governance.	Soft	Senior management focus group. Independent audit and review against Security Architecture Capability Maturity Model† for governance
Providing good stewardship and custody	Protecting other parties with whom we do business from abuse, loss of business, or personal information of value to those parties through inadequate stewardship on our part.	Soft	Independent audit, or focus groups, or satisfaction surveys
Providing investment reuse	As much as possible, the system should be designed to reuse previous investments and to ensure that new investments are reusable in the future.	Soft	Independent audit and review against Security Architecture Capability Maturity Model† of technical architecture (conceptual, logical, physical, and component)
Providing return on investment	The system should provide a return of value to the business to justify the investment made in creating and operating the system.	Hard	Financial returns and RoI indices selected in consultation with the Chief Financial Officer
		Soft	Qualitative value propositions tested by opinion surveys at senior management and boardroom level

Business attribute	Attribute explanation	Metric type	Suggested measurement approach
Reputable	The system should behave in such a way as to safeguard the business reputation of the organization.	Soft	Independent audit, or focus groups, or satisfaction surveys
		Hard	Correlation of the stock value of the organization versus publicity of system event history

*A red team review is an objective appraisal by an independent team of experts who have been briefed to think either like the user or like an opponent/attacker, whichever is appropriate to the objectives of the review.

†The type Architectural Capability Maturity Model referred to is based upon the ideas of capability maturity models.

Appendix **B**

Cultural Worldviews

HEIRARCHISTS

Marris et al. (1996) claim that hierarchists, meaning individuals whose worldview corresponds to high grid–high group, are characterized by strong group boundaries and binding prescriptions. These individuals' position in the world is defined by a set of established classifications, based on criteria such as age, gender, or race. These demarcations are considered unquestionable and are justified on the grounds that they enable harmonious life (Douglas and Wildavsky, 1982; Langford et al., 2000; Thompson et al., 1990). Hierarchical cultures emphasize the importance of establishing and preserving the "natural order" of the society. Hierarchists mostly fear things that disrupt this social order, such as social disturbance, demonstrations, and crime. Another important facet of this worldview is that people who share it show a great deal of faith in expert knowledge (Torbjorn, 2004). Hierarchical individuals trust rules and regulations and believe that institutional order and experts will be able to tackle all types of problems (Lima and Castro, 2005). Hierarchical organizations are structured according to the belief that everyone must know one's place, though that place might vary with time (Altman and Baruch, 1998). Another noticeable characteristic of members of hierarchic groups is that when they cheat, steal or overlook procedures, they operate according to the same criteria and values that apply to their formal work; they act as a group in an orderly, disciplined, and coordinated way, with respect for their own rules, limits, and precedents (Mars, 1996). Finally, hierarchists are characterized by slow adaptability to change and overdependence on regular ways of doing things (Mars, 1996).

EGALITARIANS

Egalitarians are people who can be positioned in the high group–low grid quadrant, are also characterized by high degree of the group dimension, but, contrary to hierarchists, their lives are not prescribed by role differentiation. Instead, egalitarians share the idea that individuals should negotiate their relationship with others and that no person is granted authority by virtue of his or her position (Marris et al., 1996; Langford et al., 2000). They also believe that leadership must be charismatic (Altman and Baruch, 1998). Egalitarians are characterized by an intense sense of equality; therefore, they mostly fear developments that may increase the inequalities among people. Compared with hierarchists, they tend to be skeptical of expert knowledge, because they suspect that experts and strong institutions might misuse their authority (Torbjorn, 2004). Since they dislike others making decisions about their life and actions, egalitarians prefer to have information provided to them, based upon which they can make their own personal choices (Finucane and Holup, 2005).

INDIVIDUALISTS

Individualists are people with a low group–low grid worldview. They are bound neither by group integration nor by prescribed roles, and assert that all boundaries are subject to negotiation (Karyda et al., 2005; Langford et al., 2000). They barely feel responsible for other members of society and regard the allocation of power as a matter of one's own responsibility, not dependent on position or status (Langford et al., 2000). They do not accept enforcements based on ancestry or past, since each person is responsible for oneself (Altman and Baruch, 1998). Individualists are especially concerned for the maintenance of freedom to continue life and business as usual, and they believe that carrying on through the same paths pursued thus far is the answer (Lima and Castro, 2005). They are also particularly afraid of things that might obstruct their individual freedom (Torbjorn, 2004). Mars (1996) claims that individualists are reluctant to accept rules or to follow defined instructions or procedures, especially in the cases where these appear to obstruct their current autonomy, such as, for instance, maintenance and administrative procedures and manual instructions. They tend to build short-term and instrumental relationships with their superiors. Individualism is also associated with corner cutting, rule breaking, and cheating, which means that people who share this worldview have a propensity to cheat, convert materials to their own use, short-cut procedures for ease of operation, and exploit ambiguities. When they have the choice, individualists prefer to choose short-term personal advantages over long-term corporate consequences. Individualist tendencies are also linked to a high propensity for risk taking (Mars, 1996).

FATALISTS

Fatalists are individuals with a low group–high grid worldview. They believe, like hierarchists, that their autonomy is restricted by social distinctions, but in contrast to them, they feel excluded from membership in the institutions responsible for setting the rules, and tend to see themselves as "outsiders" (Douglas and Wildavsky, 1982; Langford et al., 2000; Thompson et al., 1990). They believe that the sphere of individual autonomy is minimal and there is little room for personal negotiations (Altman and Baruch, 1998). They also believe that social classification should be based on ancestry (Altman and Baruch, 1998). Fatalists usually take little part in social life; surprisingly, they feel tied and regulated by these social groups although they do not belong to them. This fact makes this worldview quite indifferent concerning the concept of risk; what fatalists do and do not fear is mostly decided by others. These individuals would rather be unaware of dangers, since they assume that they are unavoidable anyway (Torbjorn, 2004). Concerning the type of work they prefer, most of the times, they attach themselves to jobs characterized by high degree of routine (Mars, 1996).

REFERENCE

Karyda, Maria; Kiountouzis, Evangelos; Kokolakis, Spyros; Tsohou, Aggeliki, "Formulating Information Systems Risk Management Strategies Through Cultural Theory," *Journal of Information Management and Computer Security,* Vol. 14, Issue 3, 2006.

Index

Printed in the United States
By Bookmasters